JOINTS and CRACKS IN CONCRETE

JOINTS and CRACKS IN CONCRETE

by

PETER L. CRITCHELL

THE CONCRETE LIBRARY
ADVISORY EDITOR: J. SINGLETON-GREEN

M.Sc., M.I.C.E., M.I.Struct.E., A.M.I.Mech.E.,
F.Inst.H.E., M.Soc.C.E. (France)

CR
BOOKS
(A Maclaren Company)
London

85334 011 0
'JOINTS AND CRACKS IN CONCRETE'
BY PETER L. CRITCHELL
FIRST PUBLISHED 1958 IN
THE CONCRETE LIBRARY
(ADVISORY EDITOR J. SINGLETON-GREEN
M.SC., M.I.C.E., M.I.STRUCT.E., F.INST.H.E., M.SOC.C.E. (FRANCE))
THIS REVISED EDITION
PUBLISHED 1968
BY CR BOOKS
THE MACLAREN GROUP OF COMPANIES
7 GRAPE STREET LONDON, WC2
COPYRIGHT © 1968
PETER L. CRITCHELL

REPRODUCED AND PRINTED BY
OFFSET LITHOGRAPHY IN GREAT BRITAIN
BY HOLLEN STREET PRESS LTD., SLOUGH

FOREWORD

by A. R. COLLINS, M.B.E., D.SC., PH.D., M.I.C.E., M.I.STRUCT.E.

Of all constructional materials, concrete is the one used in larger units than any other, and, it is, therefore, not surprising that the comparatively few joints that are used have difficult tasks to perform.

The wide range of structures in which concrete is employed requires a great diversity of joint designs and presents many different problems of construction. Nevertheless, joints have certain basic functions to perform, and this book will give engineers and architects a most valuable means of studying joints as joints and not merely as details of the particular range of structures with which they may be familiar.

No one could have been more suited to write this book than Mr. Critchell for he has great experience of the subject based on the firm foundation of his own research work. It is gratifying to know that his specialization will now serve to widen the knowledge of others.

Technical Director,
Cement and Concrete Association.
25 *July* 1958.

5

ACKNOWLEDGEMENTS

Although the contents of this book are largely an account of the author's own practical experience, the assembling and writing of it could not have been undertaken without the assistance and advice of Mr. John Singleton-Green.

The author is also indebted to Expandite Limited and PGH Industries Limited, Sydney, Australia, for their unreserved approval of his undertaking this work and to the many friends and associates who assisted in the production of the text and illustrations; notably Mr. G. B. Valentine for his reading of the typescript, and Messrs. R. E. Harris and B. Worner, who helped considerably with the production of the illustrations.

Grateful acknowledgements are due to the publishers of those sources of information mentioned in the text bibliography.

Lastly, the author wishes to record his appreciation of his wife's help and indulgence during the months in which his sole evening preoccupation was writing this book.

CONTENTS

7

8

LIST OF ILLUSTRATIONS

CHAPTER ONE

Introduction

As a structural material, concrete has many advantages over other materials such as wood, steel and pre-cast refractory materials, such as bricks of which various types are now being manufactured. Not only can concrete be pre-cast without requiring any heat treatment but it can also be cast in situ using appropriate formwork to produce the desired shapes.

In some respects, however, the properties of the hardened concrete lead to problems which tax the skill and ingenuity of the civil engineer in providing sound and durable structures. Whereas the strengths of steel and wood subjected to compression, tension or bending may be of the same order, the tensile and bending strengths of concrete are much lower than its compressive strength ; the ratio being approximately 1 : 9 for good quality concrete. Furthermore, concrete hardens by a chemical reaction in which water is absorbed and during this hardening process excess water added to facilitate mixing is lost by evaporation and absorption into the formwork and shrinkage occurs. This shrinkage causes tensile stresses in the concrete which can easily exceed its strength, particularly as they occur during the hardening process and therefore before the concrete has developed its nominal working strength. Thus cracks of the type shown in Fig. 1. 1 tend to develop where the stress exceeds the strength of the concrete and to overcome this tendency it is customary to articulate concrete structures by providing joints at predetermined intervals to accommodate the movement due to shrinkage and thus prevent the development of dangerously high stresses and irregular cracks.

Subsequent to the hardening process, movement continues due to moisture and temperature variations in the concrete. Increase in temperature causes expansion and increase in moisture content results in a similar but less pronounced movement. Contraction occurs when either or both the temperature and moisture content decrease.

Movements due to moisture variations will usually reduce the

11

FIG. 1. 1.   Shrinkage cracks in concrete.

effects of temperature variations, temperature increases causing loss of moisture and temperature decreases reducing evaporation. Since, however, the movements due to temperature variations are larger than those due to changes in moisture content the effect of expansion as well as of contraction must be considered and gaps must be formed in some structures to permit the concrete to expand with minimum restraint. These gaps are known as expansion joints and they differ from the joints provided to accommodate shrinkage in that they incorporate a space usually filled with a compressible material which by accommodating expansion prevents stresses developing in the concrete. A typical result of inadequate allowance being made for expansion of concrete slabs in a paving is shown in Fig. 1. 2.

A further effect of the movements due to variations in moisture content and temperature is differential movement between one part of a structure and another or through the thickness of one part. Sliding joints which form slip planes are necessary where this effect occurs, in order to prevent displacement of an adjacent section or to allow the concrete to warp.

It is necessary then to provide various types of joint in most concrete structures and in order that these joints adequately perform the functions for which they are intended it is essential that they be installed and located correctly.

When joints are installed in a structure it is essential that they do not impair the normal functions of the structure and usually it is desirable that they should blend with the general appearance. An effective black and white pattern is shown in Fig. 1. 3. In buildings both appearance and resistance to weather are important, in water-retaining structures the joints must be capable of withstanding water under pressure or in motion, and in pavings where riding quality is important, they must withstand weather conditions and preserve the smooth surface finish. Not only therefore is it important that the joints are properly located but also that the materials used to fill and seal the joints be chosen to suit the particular requirements and conditions appropriate to the structure concerned.

It is essential in the design, construction and treatment of joints, as with any other aspect of a structure, that the final details of the work be determined in advance ; if necessary based on small scale trials, to ensure that the work progresses smoothly and that the desired results are obtained. The installation of joints must necessarily be an extra operation and if not properly planned it can impede the progress of the main construction work. A small amount of care in the early stages can prevent all delays and will also result in economies and good performance.

FIG. 1. 2.   Buckling of concrete slabs due to inadequate allowance for expansion.
Note the upward displacement of the reinforcement.

Fig. 1. 3.   Expansion joint in a dairy interior. Note the staining of the painted ceiling
due to bituminous sealing compound in the ceiling.

A considerable amount of information on jointing methods and materials is available in standard works of reference, articles in trade journals and published papers and it is the intention in this book to incorporate as much as possible of this information, with the experience of the author gained through laboratory research, design and site investigations. It is hoped by this means to provide a reference book which will cover all aspects of the design and treatment of joints in all types of concrete structure. Apart from Chapter XVI which is devoted to some of the special considerations which may apply overseas, climatic conditions experienced in temperate climates such as Northern Europe are assumed.

CHAPTER TWO

The Movements to be Accommodated

Concrete is a mixture of various sizes of stone which may be crushed or in its natural state, cement and water. When water is added to the mixture of cement and stone a chemical reaction develops in which the constituents of the cement combine with the water and the whole mixture solidifies to a rocklike consistence. In practice the materials are apportioned, thoroughly mixed together and then the mixture is placed in a mould which is constructed to produce the desired shape. The final strength of the hardened concrete depends on (a) the grading, shape and quality of the aggregate, (b) the water content, (c) the cement content. The mixture must be sufficiently workable to allow it to be easily formed into the shape of the mould and then by tamping, vibration or the application of pressure to be thoroughly compacted to give the highest possible density. In order to obtain this workability it is necessary to add more water to the mixture than that required to complete the chemical reaction with the cement and this excess of water results in a number of effects two of which are :—

(1) The final strength of the concrete is reduced due to the reduction of its density.

(2) Loss of the excess water by evaporation and absorption into adjacent materials causes the concrete to shrink.

Both these effects are undesirable and it is therefore desirable to add just sufficient excess water to permit the mixture of cement and stone to be placed and compacted.

MOVEMENTS DUE TO MOISTURE CONTENT VARIATIONS

After compaction, hydration of the cement in the mixture continues and if the compacted specimen is stored immersed in water its volume will increase slightly. Continuous immersion after placing rarely, if ever, occurs in practice and although various methods are employed

17

to reduce the rate of loss of the excess water the concrete dries slowly and as it dries its volume decreases. Subsequent immersion in water will increase the volume of the concrete but will not restore it to its original value.

The movements in concrete structures which occur due to variations in moisture content may therefore be classified as follows :—

(a) SHRINKAGE—Initial contraction which occurs as the concrete sets.

(b) MOISTURE MOVEMENT*—Expansion and contraction due to absorption and loss of water respectively.

The moisture movement will always be less than the initial shrinkage by the amount of the irreversible or permanent shrinkage.

Shrinkage and moisture movements vary directly with the cement content and the water/cement ratio of a concrete mix. The type of aggregate also has some effect, sandstone and gravel aggregates giving larger movements than limestone.

The importance of allowing for shrinkage and moisture movements in a concrete structure will depend firstly on the composition of the concrete mix and secondly on the moisture conditions to which the structure is to be exposed. It will generally be found, however, that allowance for the movements due to shrinkage will be necessary in all unreinforced structures and in structures for retaining water.

EXPOSED STRUCTURES

In structures exposed to weather conditions such as bridges, pavements, buildings and sea walls the movements will be of the form shown in Fig. 2. 1, i.e., immediately after placing, initial shrinkage occurs and subsequently the concrete expands and contracts by amounts related to the moisture conditions of the atmosphere and the degree of exposure of the structure.

Rapid and comparatively large expansion and contraction effects will occur in sea walls which are washed by the sea particularly those having a southerly aspect. In bridges and to a lesser extent in buildings, contraction due to rapid drying will usually be more important, particularly as this movement will result in tensile stresses. In pavements heavy vibrating plant frequently allows a comparatively lean mix with a low water/cement ratio to be used so that both shrinkage

---

*Moisture movement is a term normally used in connection with the movement of moisture in soils and other materials by capillary and similar effects. The term is simple however, and serves very well to describe the movements occurring in concrete due to moisture variations and it is considered worth using in this connection in spite of the possibility of misinterpretation.

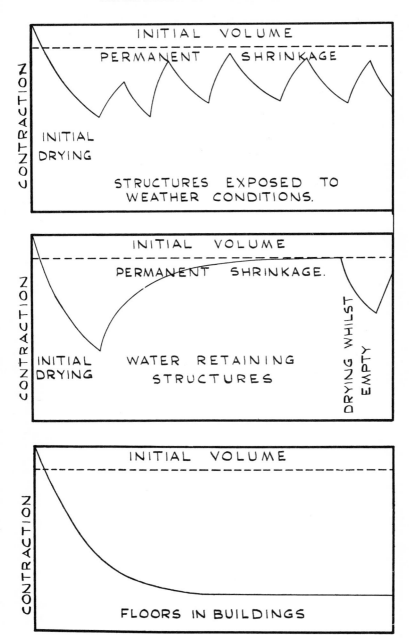

FIG. 2. 1. Moisture movements occurring in various types of structure.

19

and moisture movement effects are minimized and in general shrinkage alone need be considered.

PROTECTED STRUCTURES

In structures protected from the weather, i.e. floors in buildings, little or no moisture movement will occur after initial shrinkage as shown in Fig. 2. 1. In water-retaining structures the concrete will expand when the structure is put into service, only the irreversible shrinkage remaining except during periods when the structure is empty for cleaning or maintenance purposes and the movements therefore will be of the form shown.

SHRINKAGE VALUES

Most structures are exposed to the weather during the construction period and initial shrinkage takes place as the concrete sets and dries. The amount of this shrinkage for a conventional 1 : 2 : 4 mix made with a water/cement ratio of 0.5 will be of the order of 0.5 inch in 100 ft. Higher water/cement ratios and richer mixes will result in greater shrinkage.

Typical values for shrinkage per unit length of concrete containing different sizes of aggregates and of three different consistencies are given in Table 2. 1.[1].

In order to give some assessment of the importance of this initial shrinkage the stress which would develop in a restrained member

TABLE 2. 1. SHRINKAGE VALUES FOR VARIOUS CONCRETES[1].

Size of Aggregate (in.)	Slump (in.)	Shrinkage per unit length
	2	0.00063
$\frac{3}{4}$	4	0.00071
	6	0.00079
	2	0.00044
$1\frac{1}{2}$	4	0.00050
	6	0.00056
	2	0.00037
2	4	0.00041
	6	0.00045

would be of the order of 950 lb/sq. in. although of course in the initial stages the tensile strength of the concrete is considerably lower than this and it will crack before this stress is reached.

MOISTURE MOVEMENT VALUES

Subsequent movement due to variations in atmospheric moisture conditions will cause expansion and contraction movements to occur amounting to 0.15 to 0.2 inch or more if the restraint is small. A reasonable allowance for the permanent shrinkage will be 0.25 inch per 100 ft. except for concrete inside buildings which will usually retain most of the intial shrinkage.

THERMAL EXPANSION AND CONTRACTION

Although it is certainly true that expansion due to an increase in temperature in hardened concrete has caused structural failures it is very probable that most of the failures have occurred in adjacent sections of the structures concerned rather than in the concrete itself. In the past, emphasis has often been placed on expansion of the concrete where in order to avoid failures accommodation for contraction only was required. The use of prestressing techniques to provide a calculated compressive stress in concrete members indicates that in certain types of structure the stresses due to expansion effects may not warrant consideration and in practice they could prove to be an advantage. Tensile stresses, however, must be avoided and it is always necessary to consider contraction effects which may occur due to decreasing temperature if cracking is to be avoided.

The composition of concrete used in structures varies considerably not only in the proportioning of the materials used in its manufacture but also in the properties of these materials. As would be expected, therefore, the effect of temperature variations on concrete will depend on the mix proportions and the properties of the aggregate.

COEFFICIENTS OF EXPANSION

The dimensional changes in concrete due to temperature variations are usually expressed in terms of the linear coefficient of expansion. This is the change in length per unit length occurring due to a temperature variation of one degree Fahrenheit.

The coefficient of expansion of concrete varies principally according to the type and source of the aggregate used in its manufacture but as the coefficient of expansion of cement is higher than that of the aggregates, the proportion of cement to aggregate will also affect the coefficient of expansion of the mixture to a limited extent.

21

Gravel and quartzite aggregates give the highest coefficients of expansion in the region 0.0000065 - 0.000008/°F. whilst the lowest values in the region 0.000003 - 0.000004/°F. are given by limestone and other calcareous materials, granite and igneous rocks giving values intermediate between these two[2], [3].

The effect of the higher coefficient of expansion of cement may be more marked with those aggregates having a low coefficient viz. calcareous materials, but the variation between cement/aggregate ratios of 1 : 4½ and 1 : 7½ is rarely as much as 0.000001.

ACTUAL MOVEMENTS

The amount of thermal movement which may be expected in exposed concrete structures between summer and winter conditions that is a total variation in atmospheric temperature of 100°F., may be as high as 0.4 inch in 100 ft., but in considering the effect of temperature changes it should be remembered that the amount and rate of daily variations may be more important than the total annual variation. The maximum rate of movement probably occurs after sunset during spring or late summer under a clear sky. The maximum rate of movement measured on 60 ft. road slabs by the author under these conditions was 0.08 in/hour, the total movement occurring during the 24 hour period being 0.15 inch.

When considering the movements which occur due to variations in moisture and temperature conditions it is often necessary to bear in mind that both types of movement occur to a more marked extent at the exposed surface of the concrete than in the interior. In addition, therefore, to linear movement a distinct warping effect will also occur. This may itself give rise to stresses large enough to cause cracking when restraint due to the weight of the concrete or to adjacent sections of a structure inhibits movement.

THERMAL STRESSES

The stresses which might be developed in a fully restrained concrete slab due to thermal warping occurring during the day have been estimated by a method developed by J. Thomlinson[4] and are illustrated in Fig. 2. 2. It is evident from these values that differential movement of the concrete due to thermal effects must be given serious consideration in order to prevent cracks developing where high stresses may occur.

Attempts have often been made to overcome thermal effects in concrete roofs and pavements by laying an asphaltic surfacing material over the concrete thus to insulate it from the temperature of its surroundings. This treatment however, must be considered carefully because, in providing an insulating layer, the heat radiation and absorption properties of a black surface may in practice result in larger temperature variations in the concrete than if the layer were omitted. It is not until the layer is thicker than two inches that any advantage accrues from this method unless some means such as the application of a white wash or white chippings is made to the insulating layer of asphalt. This treatment will be referred to later.

PLASTIC FLOW OR CREEP

In some circumstances stress developing in a concrete member as a result of temperature and moisture variations may be relieved by a

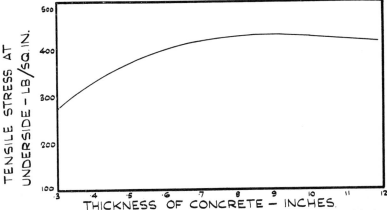

FIG. 2. 2.   Tensile stress at underside of a completely restrained concrete slab due to daytime temperature gradient.

change in volume, chiefly in the cement paste which is termed plastic flow or creep. This type of deformation is extremely complex and as it is nearly always comparatively small it is necessary to consider it in broad outline only. Plastic flow or creep is a partly elastic and partly irrecoverable change in dimension due to the application of a load to the concrete. The change in dimension which occurs depends on the magnitude of the stress induced and the time for which the load is applied. Also it varies according to the properties of the concrete and the conditions to which it is exposed. The most important variables affecting creep in concrete are as follows :—

    (1) Age. Less deformation occurs due to the application of a given load as the age of the concrete increases.

    (2) Humidity and moisture content. This is closely connected with shrinkage and moisture movement and it appears that

the creep occurring in dry conditions is larger than the sum of creep and shrinkage in moist conditions.

(3) Type of aggregate. Creep is more pronounced with sandstone and gravel aggregates than with granite and limestone.

(4) Type of cement. Larger deformations occur with Portland cement than with rapid hardening or high alumina cement.

(5) The amount of water used during mixing has a considerable effect on creep because creep occurs more particularly in the cement mortar and an increase in mixing water obviously will increase the volume occupied by cement mortar.

(6) High strength concrete exhibits less creep than weak concrete.

The amount of creep occurring due to sustained stresses in 1 : 5 Portland cement concrete using granite aggregate is shown in Fig. 2.3[5]. In practice, however, creep of 0.135 inch has been estimated[6] from the total movement of a 100 ft. slab exposed to the weather.

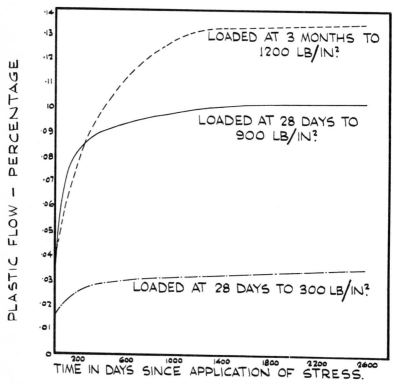

Fig. 2. 3. Typical Curves illustrating the effect of stress conditions on plastic flow.

24

Movement due to plastic flow or creep is of particular importance in reinforced structures because this property of the concrete permits stress to be transmitted to the steel without cracks developing in the concrete.

MOVEMENTS DUE TO LOADING

The most important consideration concerning movements which occur due to loading conditions is the bearing value of the subgrade. When the loading conditions imposed on the subgrade by a structure, or, when the bearing values of the subgrade vary, unequal settlement will occur and this movement can cause cracks to develop throughout the structure. This type of movement may be expected therefore in buildings of variable height and in structures carrying heavy plant and machinery. Machinery may give rise to vibration effects which must also be accommodated.

The most severe movements which have been recorded were due to settlement where structures had of necessity been constructed over unsuitable subgrades or in areas where subsidence had occurred due to mining operations. These movements were frequently of the order of several inches or, in the case of mining subsidence, of feet.

In pavements where allowance is made for movements in the plane of the structure the deflections perpendicular to this plane caused by transcient loading must also be considered. In the experience of the author relative vertical deflections up to 0.08 in. have been recorded between adjacent concrete slabs in a road on a sound foundation due to the passage of an axle load of only two tons and much larger movements than this are possible particularly when some deterioration of the subgrade or foundation has occurred.

In reinforced concrete structures the amount and distribution of reinforcement may be arranged to resist excessive movement due to loading conditions. In this case cracks frequently develop in the concrete and it is necessary that the distribution of the reinforcement be such as to limit the width of individual cracks, the sum of the widths of the cracks accommodating sufficient movement to reduce the stress in the concrete to an acceptable value.

Movements due to all causes. The relative importance of the movements which may occur in concrete structures will vary to some extent on the nature of the structure and the conditions to which it is exposed. In general the most important types of movement to be considered are shrinkage during setting and expansion and contraction due to temperature variations. Shrinkage movement during the period of setting is of great importance in all structures with the possible

25

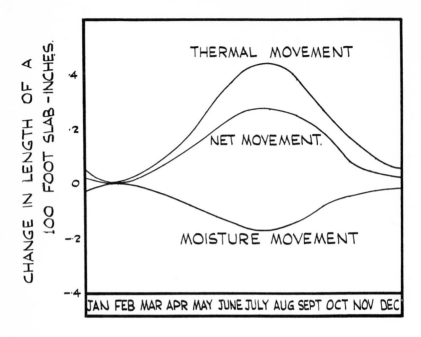

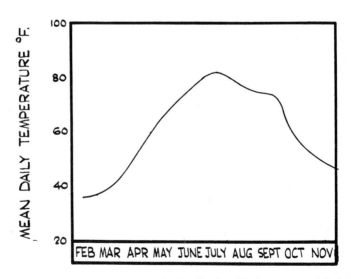

FIG. 2. 4. Average seasonal changes in length of a 100 ft. concrete slab due to temperature and moisture variations

exception of reinforced concrete roads where the use of heavy vibrating equipment permits the use of a dry lean concrete mix which will have low shrinkage. Not only is shrinkage inevitable but it results in the development of tensile stresses which concrete cannot readily withstand. These stresses may be suitably distributed to prevent serious cracking in some reinforced structures but in water-retaining structures and in all unreinforced structures it is essential that excessive stress be avoided by making provision for shrinkage movement to occur.

Shrinkage movement occurs during the construction period whilst the concrete is setting. At this time the structure is exposed to atmospheric conditions and although some degree of cover may be provided in the form of matting, wet sand or a thin membrane, temperature variations will occur in the concrete which may not be experienced during the service life of the structure. Although it may be quite unnecessary to allow for movements due to temperature variations in water-retaining structures and buried structures such as underground culverts when these structures are in service, consideration must be given to the movements which may occur during the period of construction. Typical values for the movements occurring due to temperature and moisture variations and due to plastic flow are illustrated in Fig. 2. 4[7]. Thermal expansion and contraction will occur in all exposed structures and will be particularly severe in structures which occasionally carry or impound water such as stormwater tanks, floodrelief channels, sea walls, etc. It is important in all these structures that provision be made for some movement to occur at pre-determined sections thus preventing the development of stresses which will rupture the concrete.

Movements due to temperature variations will usually be offset by contrary movements caused by loss of water by evaporation and in some instances by plastic flow but it must be remembered that plastic flow diminishes with the age of the concrete.

Movements due essentially to loading can normally be estimated with some degree of accuracy and most structures are designed on the basis of the loads they must carry when in service. Higher factors of safety should, however, be used in the design of structures such as buildings, sea walls and pavings where loads may vary significantly during the life of the structure than those used for water-retaining structures in which the load is governed by the size of the structure.

CHAPTER THREE

Crack Formation and Treatment

The main purpose of this book is to give information on the spacing, design and construction of joints in various types of structure in order to minimize random cracking. Whatever steps are taken to prevent them, however, random cracks may still develop during the life of a structure due to causes which may not have been apparent during construction or which may have arisen subsequently.

In this chapter, therefore, there will be some discussion on possible causes of random cracking and on remedial treatments which may be applied to satisfy various conditions.

The most usual cause of cracking in concrete is shrinkage during setting. In all concrete members there is restraint to shrinkage of the cement paste. This restraint may stem from adjacent members or materials, from reinforcement, or from the aggregate in the concrete itself.

In unreinforced concrete, cracking due to shrinkage or to loading may be due to:—

(1) failure of the bond between the cement paste and the aggregate

(2) fracture of the cement paste

(3) fracture of the aggregate.

The first two of these mechanisms are the most likely to occur in the case of shrinkage, due to the lack of strength and adhesive bond of the cement paste at an early age. In many cases, shrinkage cracks which develop due to restraint soon after the concrete has been placed and while its water content is high, will heal through autogenous healing in the presence of atmospheric carbon dioxide.

All of these factors militate against the exact determination of when and under what conditions concrete will crack. It has been established by various workers [8] [9] [10], however, that concrete cracks at an elastic strain of the order of 10^{-4}.

THE EFFECT OF REINFORCEMENT

In reinforced concrete the reinforcement imposes restraint on the

28

concrete because it is generally found that although the strain in the steel is equal to that in the concrete when subjected to compression, tensile stresses greater than the tensile strength of the concrete can frequently develop as a result of changes in dimension or position due to variations in temperature, moisture content or loading.

The extent to which reinforced concrete will tend to crack depends on the bond between the concrete and the reinforcement and the strain in the steel, in addition to the physical properties of the concrete itself.

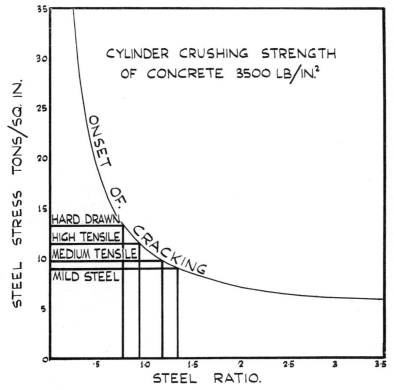

FIG. 3. 1.   The development of cracks in reinforced concrete.

The tendency for cracks to develop in a reinforced concrete structure depends on the following variables:—

(1) the strength of the concrete
(2) the bond strength between the concrete and the steel
(3) the tendency for changes in position or dimensions due to shrinkage
(4) the thickness of concrete cover to the reinforcement.

It is usually assumed that cracks will develop in reinforced concrete structures, with the exception of those designed for retaining water. In general therefore, structures are designed in such a way that the cracks are held in close contact by the reinforcement. The cracks which develop are in tensile zones where the strain in the steel is not the same as that in the concrete immediately surrounding it.

Where incorrect placing or inadequate weight of reinforcement occurs cracking at the surface is very marked and water penetration through the cracks is likely. A typical example of this defect which is unfortunately all too common in floors and similar paved areas is shown in Fig. 3. 2. In pavings it is essential that proper attention is paid to the supports for the reinforcement so that it remains near the top of the slab and is not displaced towards the bottom of the slab by construction traffic.

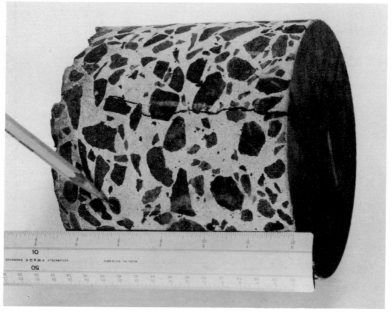

FIG. 3. 2. Shrinkage cracking due to incorrect placing or inadequate weight of reinforcement in a paving slab.
Photograph by courtesy of the Commonwealth Experimental Building Station, Sydney.

Consideration must also be given to the deflection of reinforced members under load and the influence of this deflection on other parts of the structure. In buildings brick walls are frequently erected on reinforced concrete floors and beams. Deflection of the reinforced concrete members occurs due to the loading imposed by the wall and cracking of the type shown in Fig. 3. 3 is likely to develop. In this illustration the brickwork has assumed an arch structure and typical cracking has occurred marking the arch formation. This type of failure can often be avoided by loading the concrete member with the bricks for the wall construction before starting to erect the wall. The wall is then built on the concrete member in its deflected state and no cracking develops.

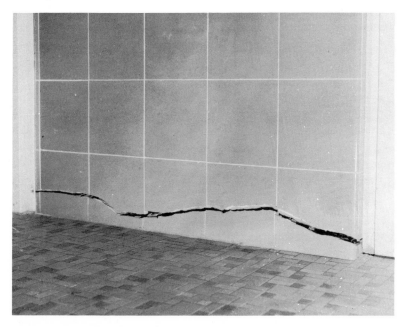

FIG. 3. 3.   Cracking of a tile-faced brick wall due to deflection of the supporting concrete slab.
Photograph by courtesy of the Commonwealth Experimental Building Station, Sydney.

In water-retaining structures and structures exposed to corrosive atmospheres it is frequently necessary to provide a high factor of safety to ensure that no cracks which may result in corrosion of the reinforcement will develop. Many methods[11] have been devised to assist in the design of reinforced concrete structures, one such[12] is a relation between the steel stress at which the concrete begins to crack, the steel/concrete ratio, and the strength of the concrete. The relation is a hyperbola of the form:—

$$t_{cr} = \left(\frac{0.05 + 2}{p}\right) u.$$

31

where t_{cr} = steel stress

  p = steel ratio

  u = cylinder crushing strength of the concrete, the tensile strength being assumed 1/12 the compressive strength.

The curve derived from this expression for concrete having a crushing strength of 3,500 lb./sq. in. is shown in Fig. 3. 1. It allows an estimation to be made of the proportion of steel of a given working stress which will result in cracks developing in the concrete. Typical permissible working stresses for various qualities of steel are shown and the associated values for the steel ratio may be used as a guide in designing a structure with freedom from cracking.

This relationship can be used only as a general guide to the onset of cracking because other variables such as the thickness of concrete covering the reinforcement and the bond strength of the concrete to the steel must be considered. Most of these variables are inter-related so that it is found that bond strength increases with thickness of cover and with the quality of the concrete, and these in turn both reduce the tendency for the concrete to crack.

Another aspect is the cracking of concrete over the reinforcement during the setting of concrete. During the setting process the wet concrete settles around the reinforcing rods and as the concrete develops its structure this settlement can cause cracks over the rods. This type of cracking occurs after a period ranging from a few minutes to a few hours after placing and is a function of the composition of the concrete and the location of the reinforcing rods. It is reported[13] to be most likely to occur with " fatty " mixes or when the cover for the reinforcement is $\frac{1}{2}$ inch or less.

Increasing the proportion of coarse aggregate to sand can minimize this type of cracking but this proportion must be limited in wet mixes otherwise differential settlement of the aggregate to the mortar can result in " map " cracking.

Crack width. The permissible maximum value for the width of cracks which develop will vary according to the degree of exposure, the cover provided for the reinforcement and the deflections due to loading. Different authorities apply different criteria[11] within the following ranges:—

Severe conditions of exposure or loading 0.01-0.09 in.

Protected or under light loading 0.05-0.09 in.

Considerable divergence of opinion exists on the limits which should be imposed and this may be expected in view of the fact that the spacing and width of cracks in identical specimens subjected to the same conditions may vary considerably even under laboratory control.

There is less likelihood of water penetrating to the reinforcement as the thickness of cover is increased, due to the diminution of crack width with depth from the surface.

Laboratory work[11] on tensile bond specimens has indicated that the width of cracks is related to the extension of the concrete and of the steel as follows:—

$$\frac{W_s}{W_c} = 1 - \frac{e_s}{e_s}$$

where W_s = width of crack at the surface of the reinforcement
 W_c = width of crack at the surface of the concrete
 e_s = extension of the embedded bar
 e_c = extension of the concrete adjacent to the bar.

This study also proved that bond strength has a big influence on the variation in the width of a crack.

The treatment of cracks. In many structural applications the treatment of cracks is unwarranted on account of the negligible widths of the cracks. In general, cracks in beams, stanchions, bridge bearers etc., are not significant but cracks in floor slabs and walls, particularly in water-retaining structures, will usually require treatment.

The treatment of cracks can take four forms:—

(a) The application of a synthetic resin adhesive which will have bond and cohesive strengths at least equal to those of the concrete.

(b) The provision of a flexible sealer which will bond to the adjacent faces of the crack and will accommodate the movement which may occur subsequently.

(c) The combination of a rigid cement mortar and a flexible sealing compound.

(d) The use of a slightly flexible pressure-resistant sealing compound and a pressure-resistant coating.

The choice of method depends on many factors. If cracking has been caused by tensile stresses which are likely to recur, the use of a strong adhesive will probably result in further cracking adjacent to the original crack, in this case the second method should be chosen. However, when cracking is due to initial shrinkage stresses and adequate allowance has been made for subsequent thermal and moisture movements, the use of thermosetting synthetic epoxy resin adhesives to repair cracks is neat, effective and economical.

Flexible sealing compounds, in general, are effective only when a substantial volume of compound is applied to a triangular or rectangular slot. The minimum dimensions for this slot should be of the order of $\frac{3}{4}$ inch wide x 1 inch deep for conventional mastics or bituminous

33

sealing compounds. The main consideration is that the movement to be accommodated by the sealing compound should not exceed 20% of the width of the sealing slot. These dimensions however can be reduced to approximately $\frac{1}{2}$ in. x $\frac{1}{2}$ in. for the more recently developed elastomeric sealing compounds such as polysulphide, polysulphide/pitch, polyurethane and silicone sealers which have extensions of 50% or more.

This means the the crack must be chased out by hand or, if it is sufficiently regular, cut by diamond or carborundum saw to provide a slot to accommodate the sealing compound. This is an expensive and tedious process and the finished repair is usually unsightly.

Synthetic resin sealing compounds are available which blend with the colour of the concrete and it is worth the extra care in cleaning by blowing out the crack with compressed air and where necessary, acid etching and washing excess acid away or by sand blasting to permit this remedial treatment to be effectively applied. There is also some indication that some of these sealers will adhere to damp surfaces. A typical gun application of an epoxy/polysulphide sealing compound to narrow joints in concrete paving is shown in Fig. 3. 4.

FIG. 3. 4. Gun application of an epoxy/polysulphide sealing compound to joints in concrete paving.

It has been shown that epoxy resins are not affected by exposure to the weather[14] except for a small degree of surface " chalking " and that epoxy/polysulphide compositions are in addition stable under immersion in water[15]. Epoxy/polysulphides have the advantage that they can be applied to damp concrete but they have lower strength characteristics than straight epoxies.

Sealing on the " dry " face. The most difficult situation to overcome is the repair of cracks on the " dry " face of a concrete wall when water is seeping through the cracks from an inaccessible " wet " face. Such situations occur in lift wells, basements and the exterior of water tanks which cannot be emptied for economic or practical reasons.

The requirements of sealing materials for sealing cracks in these situations are as follows:—

1. The material must adhere to damp surfaces
2. The material must not be water-sensitive
3. Subsequent to application, the material or sealing system must be capable of withstanding water pressure
4. In some instances the material or sealing system should be capable of accommodating movement.

Several composite systems involving the use of flexible (usually bituminous) sealing compounds and quick-setting cement mortar have been described elsewhere[16]. Typical systems are shown in Fig. 3. 5[17].

When it is not necessary to accommodate movement, cracks through which water is seeping may be effectively sealed by forming a small triangular slot and caulking into this slot a cementitious/asbestos fibre caulking compound such as those used for caulking joints in cast iron and concrete pipes.

FIG. 3. 5. Typical systems for sealing joints against reverse water pressure.

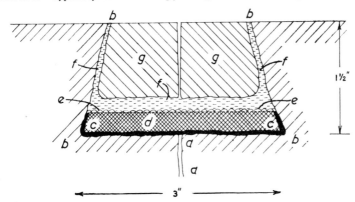

aa	original crack
bbbb	dovetailed chase
cc	bituminous primer
d	bituminous hot poured blanket
ee	coarse clean grit set in the hot bitumen
fff	1/1 sand/cement adhesion coat
gg	$2\frac{1}{2}/1$ sand/cement filling with a plane of weakness opposite the original crack.

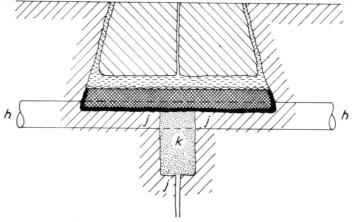

hh reinforcement can pass through the chase
jj small chase cut out $\frac{1}{2}''$ wide into the crack
k   temporary sealing with accelerated neat
    Portland cement to give a dry chase.

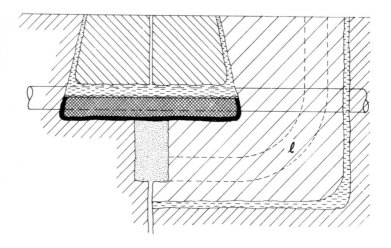

Frequently it is necessary not only to seal isolated cracks but also general cracked areas and areas of porous concrete due to segregation or indequate compaction. In addition therefore, to the crack sealing operation a surface treatment may be required.

When this treatment is required on the water face it is best achieved with a tar, bitumen, rubber/bitumen membrane in the form of an emulsion or a tar/epoxy of a polyamide cured system which will " wet " and penetrate damp surfaces.

36

In the event of the surface treatment being required on the "dry" face it is necessary either to progressively seal the surface using an accelerated cement mortar render, localizing the leak to a predetermined point and then, when the rendering has hardened, sealing this point with a "flash-setting" low water content cement paste, or first to stop seepage by one of the other methods described in this chapter and then apply a cementitious paint which expands on setting and thus withstands high "reverse" pressures or a polyamide cured tar/epoxy with an appropriate penetrating primer.

Spalled joints. In addition to their application to cracks in concrete, epoxy/polysulphide adhesives are particularly suitable as concrete adhesives. Thus, spalled joints may be repaired using a dry-mix mortar and an epoxy/polysulphide adhesive, the procedure being as follows[18]:—

1. Clean and dry the concrete surface and remove any laitance or loose material with a wire brush.
2. Mix the required amount of epoxy concrete adhesive, and thin with about 10 per cent by volume of acetone.
3. Apply the thinned epoxy concrete adhesive by brush or rubber squeegee and spread the film as thinly as possible.
4. Allow 15 minutes at least for the acetone to evaporate and then sprinkle a little dry cement on the surface.
5. Place the new concrete or cement mortar up to the adhesive.

The pot life of these adhesives varies and the permissible period between application of the adhesive and the placing of the concrete may vary from $\frac{1}{2}$ hour to three or more hours depending on composition.

Construction joints. The most frequent cause of cracking in concrete is loss of bond between the successive pours of concrete at construction joints.

This cracking is promoted by inadequate spacing of contraction or contraction/expansion joints but obviously is most seriously affected by the method adopted in making the joint. In general, however, it is considered to be expedient to incorporate a suitable "waterbar" in all construction joints where water penetration is a hazard. The cost of such an installation compared with the supervision necessary to achieve a waterproof construction joint and the difficulty of the task of dealing with any leakage that may occur subsequently is insignificant.

Several methods of making construction joints have been adopted and many methods, supported by correct design and the skill and care of the operators have been consistently successful. Among these are the following:—

(1) Hosing the surface of the concrete already placed while it is still " green ", thus exposing the aggregate.

(2) Acid-etching the surface of the concrete already placed after it has hardened and then washing down with water.

(3) Applying a synthetic resin concrete adhesive to the cleaned surface of the concrete already placed. Polyvinylacetate is suitable except in water-retaining structures where a poly-sulphide/epoxy should be used.

(4) Exposure of the aggregate in the surface of the concrete already poured by mechanical means such as sand blasting, " sparrow-picking " or " bush hammering ".

The choice of method is not so important as the care devoted to its execution. All of the methods require the conscientious application of the operators, not only in preparing the surface but also in ensuring that the surface is not contaminated after treatment. In this regard sliding forms have definite advantages.

The essence of success in all concrete work is attention to detail and this calls for the co-operation of all trades and professions connected with the construction work.

CHAPTER FOUR

Types of Joint

In order to limit or prevent the stresses which develop as a result of tendencies for volumetric changes or changes in position joints are provided in concrete structures. Many types of joint have been devised to accommodate the various movements which may occur in a structure and in addition to these other joints must be made whenever there is a break in the construction programme. It is convenient therefore to consider the various types of joint in two groups:—

(a) Functional joints which are installed to accommodate movement.

(b) Construction joints which are made when there is a break in the construction programme.

FUNCTIONAL JOINTS

Functional joints may be installed in a concrete structure to accommodate six types of movement as follows :—

1. *Shrinkage during setting.* In order to prevent random cracks developing in a structure as a result of the stresses induced by shrinkage, complete breaks or planes of weakness may be formed during construction. These planes of weakness are known as contraction joints and may be in three forms.

(a) Complete contraction joint (Fig. 4. 1). In this type of joint the bond between adjacent sections of a structure may be broken completely by painting one face with a bituminous material or by setting a layer of waterproof paper or roofing felt against the face of one section before casting the next section up to it.

(b) Partial contraction joint (Fig. 4. 2). When structural stability is required between sections of a reinforced concrete structure separated by a contraction joint it is sometimes convenient to continue the reinforcement across the joint. Due to the

39

presence of reinforcement the movement at these partial contraction joints is usually very small.

(c) Dummy joints (Fig. 4. 3). "Dummy" type contraction joints are used more particularly in thin sections of concrete. In these joints a plane of weakness is created by forming a groove in either or each of the surfaces of the concrete, the total depth of groove being one third to one fifth of the thickness of the section. Dummy joints function by cracking when sufficient stress has been developed to overcome plastic flow and to crack the member where the thickness of the section has been reduced. They are used mostly in pavings because the plane of weakness can conveniently be formed either by setting a suitable former across the formation in advance or by inserting a blade and former during the laying operation.

2. *Expansion* (Fig. 4. 4). Allowance for expansion due to temperature effects must frequently be made in structures where one or two dimensions are large. In such structures gaps must be formed at predetermined intervals to allow adjacent sections to expand with minimum restraint. These gaps are usually formed by setting a strip of compressible material against the face of one section before the second section is cast up to it and leaving it embedded in the concrete. As the main function of the gaps thus formed is to accommodate expansion of the concrete, they are called Expansion Joints.

The compressible material used to form and subsequently to fill expansion joints is known as the joint filler. The functions and properties of joint fillers are described in Chapter Six.

Shrinkage of the concrete is also accommodated to some extent by expansion joints although, as will be seen later, the spacing of expansion joints is not as a rule adequate to prevent dangerous stresses from being developed due to shrinkage effects.

3. *Sliding and warping.* When variations in temperature, moisture content or loading result in a tendency for one part of a structure to move in a plane at right angles to the plane of another part it is necessary to provide a slip plane between the two parts thus enabling freedom of movement in both planes. Three types of movement may give rise to the necessity for this type of joint namely :—

(i) Variations in length due to temperature or moisture changes occurring at right angles to similar variations in another part of the structure. This type of movement occurs where the roof of a building or a covered water reservoir rests on the walls.

FIG. 4. 1.   A complete contraction joint under construction.

FIG. 4. 2. A partial contraction joint incorporating a waterbar.

FIG. 4. 3.   A sawn " dummy " joint.

FIG. 4. 4.   An expansion joint being constructed between double columns.

(ii) "Warping" due to moisture and temperature gradients through the thickness of a section. The longitudinal joint in a long paving is an example of a joint to accommodate warping effects.

4. *Deflections due to loading.* This type of movement may occur between columns supporting a roof and the floor slabs of a structure and a sliding joint is required between the column base and the floor to provide freedom of movement and avoid excessive stresses developing in the floor slabs.

Sliding and warping joints are usually formed either by applying a layer of bituminous material to one of the surfaces before the other is cast up to it or by setting a layer of waterproof paper or roofing felt between the surfaces.

5. *Vibration.* Occasionally, joints which are designed principally to accommodate sliding movement incorporate a joint filler of the type normally used in expansion joints. In such joints allowance is thus made for expansion movement or as is the case in machine mountings the joint filler insulates the floor from the vibrations of the machinery. construction joint is shown in Fig. 4.5.

CONSTRUCTION JOINTS

As the name implies, construction joints are used to simplify the construction of a structure. Pure construction joints are not intended to accommodate movement and in fact, every effort is directed towards preventing movement from occurring at these joints. However, although the greatest possible care may be taken to obtain a good bond between abutting sections of concrete it is frequently found that cracks develop at these joints as a result of stresses arising from variations in temperature, moisture content, or loading. It is therefore most desirable to arrange the construction programme in such a way that when breaks are necessary at the end of a working period or due to complicated shuttering, or to some other cause, a functional joint may be formed. The joint may then be designed to accommodate movement, incorporating the appropriate materials for this purpose and to prevent the ingress of water or foreign matter which may detract from the good performance or appearance of the structure.

It frequently happens, however, when large volumes of concrete are being placed that it is not convenient or economic to form functional joints every time there is a break in the constructional work. Probably the most common type of construction joint is the lift joint which is made between successive layers of concrete placed during the construction of a wall. The use of shuttering which may be raised continuously as the concrete sets and accelerated setting achieved by the addition of quick-setting agents to the concrete mix, or by the application of a vacuum to extract excess water may overcome the necessity for construction joints, and such methods will in many cases be justified by increased production and better performance.

When construction joints cannot be avoided they should be made in the following manner.

1. The surface of the concrete which has already been placed should be thoroughly roughened by bush-hammering if it has hardened, or by wire brushing or hosing down if still green, so as to expose the aggregate and to remove the layer of high fines content concrete which is often found at the top of a lift. If the concrete is compacted using vibrating equipment vibration should not be continued long enough to promote the formation of a thick layer of high fines content concrete at the top of the layer of concrete being placed. The consistence and grading of the concrete should also be chosen to minimize this effect.

2. The surface of the concrete should then be cleaned thoroughly to remove all foreign matter such as waste and sawdust from shuttering work. This is best done by hosing down with water but this method may not always be possible, in which case thorough brushing or the use of a jet of compressed air is necessary.

3. If not hosed down during cleaning the surface should be thoroughly wetted.

4. When all the pockets of surface water have soaked into the concrete a layer of mortar similar in proportions to the matrix of the concrete should be worked into and spread over the surface.

5. Before this mortar has set the next layer of concrete should be placed and tamped into the mortar so that the two mixtures blend together. Particular care should be taken at construction joints to ensure that water or cement matrix does not leak between the shuttering and the concrete already in position.

Construction joints should always be deliberately formed using appropriate shuttering in order to form a plane surface to which the next layer of concrete is to be laid. Under no circumstances should the concrete be allowed to find its own level when a vertical construction joint is required. Stop end boards should be set against the shuttering forming a vertical construction joint which should subsequently be treated in the same way as a horizontal lift joint. A well prepared construction joint is shown in Fig. 4.5.

FIG. 4. 5.   A correctly prepared construction joint.

Construction joints are nearly always the weakest points in a structure and, in addition to exercising particular care in making these joints, it is frequently desirable to adopt special methods for sealing and camouflaging them so as to ensure that the performance and appearance of the structure are not impaired.

It is frequently necessary therefore to install some type of jointing material in construction joints either during or subsequent to making the joints. When jointing materials are used the design and method of forming the joints, whether they be construction or functional joints, must be chosen to suit the types of material which will perform the function required. Before considering the design and construction of the various types of joint in detail some knowledge of jointing materials is required. It is appropriate therefore to pass to other aspects and return to joint design in a later chapter.

CHAPTER FIVE

Joint Spacing

There is much divergence of opinion on the most effective and convenient spacings for the joints which may be considered necessary in various types of structure. There are indeed, many engineers who do not consider joints to be necessary at all and structures of various types have been built without joints in some cases with complete success.

The satisfactory performance of these structures is probably due to extremely robust design and scrupulous control over all the construction processes. It is not often possible from economic considerations, however, to adopt such extravagant designs and employment of staff so that for every one successful structure not incorporating joints there are many similar structures where deformation and cracking have resulted in severe disfigurement or poor performance. In view of the total capital expenditure involved and the high cost of possible remedial treatment it is usually expedient to devote the small amount of time, labour and expenditure to the provision of correctly designed joints at the appropriate spacing to minimize the possibility of cracks developing due to movement or poor workmanship.

The spacing of joints in structures must be varied according to the stresses which may develop in the concrete. These stresses depend on the degree of exposure of the structure and the restraint imposed on its members due to their construction or to their position in relation to other parts of the structure and to the natural ground formation. Joint spacings will therefore be considered in relation to the various types of structure, the exposure conditions and method of construction.

BUILDINGS

The varying exposure to which the parts of a building are subjected results in a complicated distribution of stresses. Considering these stresses alone it would appear that joints should be

D

installed at a closer spacing in the roof than in the walls and that an even larger spacing could be applied to the floors which are completely protected from the weather.   Whereas it is possible to do this, however, the adoption of staggered joints frequently results in sympathetic cracking due to mechanical bond between adjacent jointed and un-jointed sections.   Large buildings are usually constructed, therefore, in sections of predetermined length which are completely separated from each other by expansion joints.   Joints to accommodate shrinkage or contraction are not normally required in addition to expansion joints, except possibly in floors, because structural concrete is always reinforced.   The suggested spacings for joints in buildings are given in Table 5.1, some allowance being made for the effect of time of construction because this will determine whether expansion or con-traction movement will predominate subsequently.

These spacings are suggested on the basis of movements resulting from normal exposure.   It is quite common, however, that somewhat different spacings may be more convenient to the structural design of a building or may be dictated by other contingencies such as the need to allow for varying loading conditions or for unequal settlement due to a variable subgrade.

It is generally considered desirable to isolate the floors from the external walls of large buildings.   In addition the floors may be con-structed in small bays with contraction and possibly expansion joints at spacings similar to those suggested for unreinforced pavements in order to prevent unsightly cracks developing at construction joints. When this procedure is adopted floors should be constructed in bays which are as near to square as possible so that the movement due to shrinkage is the same at all joints.   In most buildings it is not necessary to provide space for expansion in the floor except to isolate the floor from the walls, or from other parts of the building such as lift shafts which subject the subgrade to different loading conditions.

The most convenient and therefore most common method of installing joints in a large building is to divide the structure into a number of wings, joints being provided between the wings and the main part of the building.   A typical example of this design is shown in Fig.5.1.

WATER-RETAINING STRUCTURES

In water-retaining structures the surface of the concrete is usually exposed to water pressure which may be due to the static head of water alone or may in addition result from motion of the water.   As the function of these structures is to retain water, particular care is

TABLE 5. 1. SUGGESTED SPACING OF EXPANSION AND CONTRACTION JOINTS IN VARIOUS TYPES OF STRUCTURE.

		Reinforced Structures			Unreinforced Structures		
		Buildings	Water-retaining and Sea Walls	Pavements	Water Retaining	Sea Walls	Pavements
Maximum expansion joint spacing (ft.)	Summer construction	100	90	150	80	60	100
	Winter construction	80	70	90	60	50	60
Maximum contraction joint spacing (ft.)	Summer construction	—	30	—	20	20	20
	Winter construction	—	35	—	20	25	20

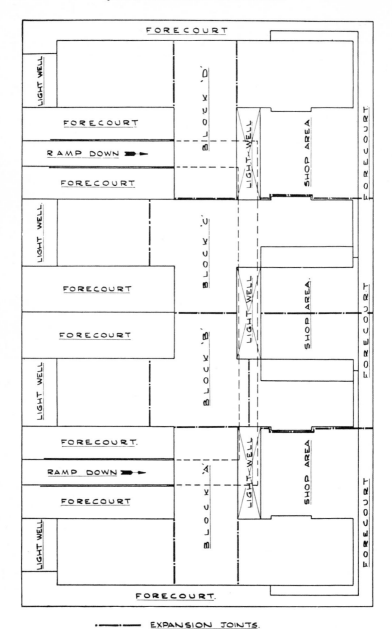

EXPANSION JOINTS.

FIG. 5. 1.   The location of expansion joints.   A typical design in which a building may be constructed in blocks separated by expansion joints.

necessary in selecting the appropriate spacing of joints. It is essential to prevent cracks developing through which water under pressure may penetrate either through the structure or to the reinforcement.

Considering conditions of exposure *alone*, smaller movements would be expected in water-retaining structures than in other types of structure because water-retaining structures are usually insulated from climatic variations by :—

    (a) the water they contain,

and (b) the back-filling of soil which is banked against the walls and in the case of covered structures applied to the roof.

Considering the movements which may occur whilst a water-retaining structure is in service it might be supposed that the need for providing joints in these structures does not exist. These structures are, however, under construction for a number of months and during this time they are fully exposed to the weather. It is during this period that cracks may occur due to shrinkage and moisture or thermal movement. If joints are not deliberately provided to accommodate this movement the planes of weakness which usually exist at construction joints will fracture and these joints will then act as functional joints. Unless some precautionary measure has been taken to seal construction joints during construction these joints will then provide an easy path for water to pass through the structure when it is put into service. The use of construction joints incorporating a sealing device forms the basis of a method of design and construction which is often used in water-retaining structures which are designed as heavily reinforced monoliths. It is not possible in such structures to provide joints which produce complete breaks, and thus allow freedom of movement between two sections, because this destroys the structural stability of the structure. The joints provided in monolithic structures should therefore be of the partial contraction type embodying a sealing device and these should be provided, at least, at the spacings suggested for complete contraction joints. Preferably a somewhat closer spacing should be employed to ensure that cracks, if any, will occur at these joints because the strengthening provided by the reinforcement will make the joints approach the strength of the rest of the structure.

Expansion joints, of course, will be omitted so that expansion of the concrete will result in compressive stresses and as the reinforcement will normally be sufficient to prevent buckling. These stresses will serve to strengthen the structure by preventing the development of tensile zones.

Reinforced structures not designed on the monolithic principle and mass concrete structures will normally be provided with both contraction and expansion joints.

The contraction joints in these structures may be complete or partial where reinforcement is used but expansion joints must always be complete breaks between adjacent sections. The suggested spacings for the joints in water-retaining structures should be used as a general guide and should give satisfactory results under the construction and service conditions of reservoirs, sewage tanks and channels and culverts which are in continuous use.

Flood relief channels, aqueducts, retaining walls, large storm water tanks and other structures which hold or carry water only periodically may require somewhat closer spacings than those suggested in order to accommodate larger and more rapid movements occuring due to complete exposure. This is particularly true of sea walls, especially those on a southern shore which are washed by the sea. There is, indeed, a very good case for the use of pre-cast blocks in the construction of sea walls, not only from the standpoint of simplifying construction, which is an all-important factor in emergency measures, but also because movements are uniformly distributed. This has an important bearing on joint sealing compounds which will be considered later.

PAVINGS

The spacing of joints and in fact the types of joints considered necessary in concrete pavings is probably one of the most controversial topics in concrete pavement design.

One of the most important requirements for concrete pavings, such as roads and runways, which carry vehicular traffic is a smooth riding surface. The irregularities caused by joints in these pavings have in the past led to serious criticism of concrete as a construction material. The tendency therefore has been to increase the spacings of joints so as to reduce the number of irregularities. One of the longest experimental continuously reinforced pavings was laid on U.S. 22 near Hamburg, Pennsylvania in 1957[19]. Fig. 5.2. A length of 10,800 feet of 24 ft. wide carriageway was laid without joints of any type, rectangular mesh reinforcement being incorporated to control cracking. According to past experience on other roads of similar design minute hair cracks are anticipated at approximately six foot spacing. Laboratory tests had indicated that the width of these cracks could be restricted to 0.025 inch by incorporating longitudinal reinforcement at the rate of 0.5 per cent of the cross-sectional area of the pavement. The reinforcement used in the project was thus between 13 and 16 pounds per square yard for slab thicknesses of seven to nine inches. Obviously no conclusions can yet be drawn

(Photograph by courtesy of Contractors and Engineers Monthly, New York)

FIG. 5. 2. A continuously reinforced concrete road in America.

from this work except that construction was greatly simplified and a smooth riding surface was obtained.

Reduction of the frequency and width of joints in pavings, however, has resulted in structural failures. These failures have been caused not only by shrinkage effects but also by expansion, which produces serious fractures and deformation involving complete removal and reinstatement of the affected area.

Considerable research and development work has therefore been devoted to methods of forming joints in pavings with the object of reducing the surface irregularity in the region of the joints. New techniques in forming and finishing joints in pavings have reduced surface irregularities to a level comparable[20] [21] with that achieved in machine laid bituminous surfacings which are generally considered to be quite acceptable from the aspect of riding quality. Whereas, therefore, the maximum spacings are given in the Table it may well be found that considerable reduction of these spacings with the appropriate reductions in joint widths will improve riding quality further.

In the light of present experience it would appear that, in common with other types of structure, pavings may be constructed in two ways as follows :—

(i) Monolithic with relatively heavy reinforcement to limit the width of cracks which will be at a very close spacing.

(ii) Unreinforced or lightly reinforced with freedom of movement provided by joints at pre-determined intervals.

It is certainly true that these two methods may be used in the construction of comparatively small paved areas but other factors such as loading effects and effects on adjoining structures may then arise so that isolation or sliding joints may be required. For example, in garage forecourts an isolating expansion joint is necessary between the petrol pump installations and the paving which should also be provided with expansion space around its perimeter unless it is unrestricted. If reinforcement heavier than 10 lb./sq. yd. is provided no other joints may be necessary apart, possibly, from construction joints.

Runways. The spacing of joints in large paved areas such as runways should be the same in both directions, thus dividing the paving into square slabs. Runways in particular are usually constructed in plain concrete so that both expansion and contraction joints may then be chosen as the width laid by the machinery used for placing and finishing the concrete.

Roads. When transverse joints are installed in roads they should be continued through the kerbs in order to prevent sympathetic cracking in the sections of kerb bridging the joints. It is neither necessary nor desirable to provide an expansion joint between the road slabs and the kerb.

It is generally recommended that in addition to the transverse joints longitudinal warping joints be provided in roads wider than 15-16 ft. The longitudinal joint in a road is necessary to accommodate:—

(a) Unequal settlement of the subgrade.

(b) Differential moisture changes in the soil.

(c) Warping of the concrete due to moisture and temperature gradients.

Construction Joint Spacing

As construction joints are provided to simplify construction, or even to make it possible, the spacing of these joints is determined by the type of work, the site conditions and the production capacity of the plant or labour employed. It is, however, highly desirable that consideration be given to the need for construction joints early in the design or construction of a structure so that they may be reduced to a minimum by appropriate spacings of functional joints. It is also important from economic considerations that the spacing of construction joints be planned such that the joints, whilst complying with production capacity, do not occur where they may not be properly constructed or are undesirable because of stress conditions.

CHAPTER SIX

Jointing Materials — The Joint Filler

Although it would be reasonable to use the term joint filler to describe any material which is used to fill joints, no single material has yet been found which is sufficiently compressible to accommodate the expansion movement whilst providing an adequate seal against the ingress of water and foreign matter. In the various types of concrete structure to be considered three different materials may be necessary to satisfy the requirements of the joints to be installed. These materials can conveniently be put into three classes according to their functions and properties as follows :—

(1) The joint filler. This is a term which may be applied to the strip of compressible material used to form and to fill the expansion joints in a structure.

(2) The sealing compound. This is the material which usually has a plastic consistency and is applied to the joint in the form of a liquid, paste or preformed strip, its function being to prevent the ingress of water or foreign matter.

(3) The waterbar. This term is applied to a strip of material which is set across the joint in the concrete during construction so as to form a water-resisting diaphragm.

The functions of the joint filler

The definition of the term joint filler as applied to concrete pavings is given in B.S.S. 2499 as follows :—

" Joint filler. The strip of compressible material used as a spacer between two concrete paving slabs permitting the slabs to expand as a result of thermal and other changes without developing deleterious compressive stresses in the concrete."

Whereas, however, a joint filler should, and usually does, prevent the development of deleterious compressive stresses it is unlikely that failures due to expansion, such as that illustrated in Fig. 1. 2 have been due purely to compressive stress. Due to the temperature gradient

58

through the thickness of the concrete the amount of expansion will vary. Warping will then occur causing flexural or tensile stresses and the concrete will rupture earlier due to these stresses than due to pure compression.

The functions of the joint filler are therefore as follows :—

(a) To form the expansion space during construction.

(b) To allow the concrete to expand freely.

(c) To support the sealing compound.

In order to fulfill these functions the joint filler must have the following properties :—

(i) Compressibility without extrusion, i.e. it must be cellular.

(ii) The ability to recover as nearly as possible to its original thickness when pressure is released.

(iii) Durability and resistance to rotting.

(iv) Sufficient rigidity during handling and placing to permit the formation of a straight joint.

Some limit must be set to the compressibility of the joint filler in order to prevent its width and thus the width of the joint from being reduced during the placing and compaction of the concrete. The limits set by the American Federal Specification No. HHF 341 are that a pressure of not more than 500 lb./sq. in. shall compress a specimen $4\frac{1}{2}$ in. x $4\frac{1}{2}$ in. to half its original thickness and whereas this Specification applies to pavings only, this limit is probably reasonable for other types of structure.

TYPES OF JOINT FILLER

It is the intention to consider under this heading all the materials known to have been used as joint fillers although some of these do not possess all the properties already described.

Bitumen. This was probably the first material to be used to fill expansion joints and it is still used occasionally although it is not entirely suitable.

Bitumen may be used in two ways. It may be placed during construction as a preformed board usually enclosed by layers of felt, sufficient fine mineral or vegetable matter being incorporated in a hard grade of bitumen to inhibit flow. It may also be poured or gunned as a liquid or paste into a cavity formed during construction by a removable former.

The main advantage sometimes quoted in favour of bitumen joint fillers is that they provide an effective seal against the ingress of water and grit. Whereas this advantage may exist during warm weather, when adjacent sections of concrete are expanding, the effect of this expansion is to cause the bitumen to extrude from the joint.

Fig 6.1. Extrusion of a cork filled bitumen joint filler.

If, as is likely, expansion is followed by contraction the bitumen being hard is unable to accommodate the movement without fracturing or failing in adhesion to the walls of the joint.

Where only a small amount of expansion movement is anticipated as for example in buried water-retaining structures one or more layers of roofing felt have been used as a joint filler. The performance of roofing felt is similar to bitumen board but as it is used only in narrow joints accommodating negligible movement, extrusion is not usually a problem.

Bitumen containing cellular materials. In an attempt to overcome extrusion cellular materials in granular form have been incorporated in preformed bitumen joint fillers. Granulated cork is a typical material for this purpose. The properties of these joint fillers vary according to the quantity of bitumen they contain. When this type of joint filler is dense and impermeable, bitumen fills the voids between the granules. Although, due to their cellular nature the granules may be compressed initially by the expansion of the concrete adjacent to the joint they do not remain in this stressed condition but expand to their original size causing the bitumen to extrude from the joint. A typical example of this effect is shown in Fig. 6. 1.

The performance of these materials is therefore similar to that of bitumen joint fillers.

If only sufficient bitumen is used to bind the granules together somewhat better performance is obtained and due to its similarity with other materials this type of filler is considered under the next heading.

Cork. Natural cork has been used as a joint filler in the form of strips and in the form of granules which may be applied loose or in the form of a board in which the granules have been coated with a suitable binding medium.

Cork strips. Natural cork can be obtained in the form of a board made by cementing strips of cork together. Whilst being very resilient and compressible, however, this material is prone to damage during storage and handling. After a few years in a joint the strips separate leaving gaps into which the sealing compound may be forced by pressure of water, the action of traffic or gravitational flow. The board may be coated with bitumen, presumably to prevent rotting, but such treatment is unnecessary.

Cork granules. Cork has been used in the form of granules bound with rubber, resin or bitumen. All three types of material may be expected to possess the properties considered desirable in joint fillers, providing that, in the case of the bitumen-bound material only sufficient bitumen is used to bind the cork granules together.

These materials will generally be porous and will absorb water and this may result in swelling.

Loose granules. Cork granules may also be used for maintaining horizontal joints, such as those in pavings or floors, but there is a tendency for loose materials of this type to work under the adjacent slabs of concrete. Moreover, being extremely light they are difficult to handle on exposed sites during windy weather.

Dehydrated cork. This material has been developed in the United States of America specifically as a joint filler. It is packaged in air-tight wrappers and when the wrapping is removed it absorbs moisture very readily showing a considerable increase in volume. As a joint filler it absorbs moisture from the concrete placed against it and tends to expand so that it is likely that it is always in a state of compression when the concrete has hardened. It appears likely that this type of joint filler will adequately seal joints in continuous immersion conditions but they tend to deteriorate under continuous atmospheric exposure. Thus it would seem that this class of material can be considered only as a particularly efficient joint filler associated with a sealing compound except possibly in water-retaining structures.

Wood. Wood, either in its natural state or in the form of a board manufactured from pulp or wood chips, is probably the most widely used joint filler. In one or more of these forms it will generally be found not only economical but also efficient in fulfilling the functions of a joint filler.

Natural wood. There are many varieties of wood which are readily compressible and possess sufficient recovery to meet the requirements of a joint filler. The following are some of the varieties which have been used :—

Cypress	Obeche
Redwood	Deal
Spruce	Various white pines

Soft woods such as these are easily worked and are sufficiently rigid to withstand the rigorous conditions to which joint fillers are exposed when heavy equipment is used for placing and compacting the concrete. They will all absorb water and may swell in the process. The amount of recovery after compression will diminish with repeated loading and with moisture content. Only knot-free wood should be used as a joint filler in order to ensure uniform compressibility.

Rot-proofing treatments are often applied to wood joint fillers the most effective treatments being pressure impregnation with creosote or a pentachlorophenol. It has however, been reported[22] that untreated soft wood has been found in a good condition after 20 years in a road expansion joint.

Wood pulp board. Several types of soft board manufactured from wood pulp have been used as joint fillers. Whilst being readily compressible without extruding many of these boards lack recovery. They generally absorb water and swell considerably. One material in this class which has been developed to satisfy the requirements of an expansion joint filler contains a high proportion of sugar cane fibre and is known as fibreboard.

Fibreboard. Fibreboard joint fillers usually contain up to 70 per cent of a resilient fibre such as cane and thus have very good recovery after compression. They are normally impregnated with bitumen for rotproofing purposes and may also be treated with an insecticide to discourage attack by termites.

Fibreboard joint fillers do not extrude during compression but they should not be compressed to such an extent that they become dense because this will displace the impregnant. Exuded impregnant will not only stain the concrete but may flux the sealing compound rendering it ineffective. The reduction in thickness of the board under compression should not exceed 50 per cent in order to avoid this effect.

These joint fillers may also be caulked whilst in the joint so as to fill the gap caused by shrinkage of the concrete during setting and thus to provide complete support for the sealing compound during application.

Chip board. A board composed of small wood chips bound together by resin has recently been used as a joint filler. Because of its rigidity, this material has been found suitable for forming expansion joints in pavings by setting it up on the formation and laying two or more consecutive concrete slabs continuously, using heavy placing and vibrating equipment.

Due to their method of manufacture these fillers are usually dense. As the wood chips are coated with resin these fillers do not initially absorb water. They are not as a rule readily compressible and do not recover completely after repeated or heavy loading. It is possible that this type of filler could be rendered more efficient in service by the use of a binding medium which possesses a small amount of flexibility.

Sawdust. Sawdust or loose wood chips have been used to maintain expansion joints in pavings when the original joint filler had become displaced or had disintegrated. This material however has the same disadvantages as cork granules, namely a tendency to work underneath the concrete slabs and difficulty in handling in windy weather or on exposed sites.

Rubber. Its high compressibility and recovery make rubber an obvious choice as a joint filler. As a result natural and synthetic rubber have both been used in various forms to fill joints in nearly all types of structure.

Natural rubber. A compounded rubber is normally used for the manufacture of joint fillers in order to give enhanced properties of recovery and resistance to weathering. In solid form rubber extrudes under compression but on account of the compressibility of the material and the effectiveness of the seal it provides when maintained in the compressed state solid rubber may be used as a joint filler. Typical examples of its use are bearing and hinge joints in bridges and in spigot and socket or flanged pipe joints.

Cellular Rubber. A joint filler of more general application is cellular rubber. This material is usually a vulcanized rubber which has been expanded either by the injection of an inert gas or by the generation of gas within the rubber by chemical means.

Cellular rubber is usually readily compressible and has good recovery after compression. It is however inclined to be too pliable in some applications and may even be reduced in thickness by the weight of the concrete placed against it thus reducing the width of the expansion space.

When used in pavings cellular rubber is difficult to place and in service, tends to work up out of the joint, presumably due to vertical deflections of the slab edges under the transient loading of the traffic. This effect occurring in joints incorporating dowel bars for load transfer between adjacent slabs is illustrated in Fig. 6. 2. It is evident that this is a very positive effect because the joint filler has been forced through the sealing compound, which, in this joint, was a relatively hard rubber/bitumen composition.

When the cells in cellular rubber joint fillers are not inter-connected no absorption of water occurs but water absorption and swelling may be very marked if some connection between the cells does exist.

Expanded plastics. A wide range of cellular materials is now being manufactured using plastics or synthetic rubbers. These are usually polymerized resins which may be modified by the introduction of plasticizing agents to obtain the desired stiffness or by the incorporation of another polymer to produce surface " tack." The latter modification is particularly promising in achieving a non-extruding joint filler which will also seal the joint because a readily deformable basic material may be chosen which will result in a very low bond stress developing when movement is accommodated.

Although only limited experience has been obtained on the performance of these joint fillers the inert nature of many of the

FIG. 6. 2. Cellular rubber working out of a dowelled expansion joint.

E

polymers used supports the belief that they would have very good weathering properties.

One problem which exists at the present time is the production of these materials in boards large enough to make installation an economic proposition in general civil engineering work. A possible solution to this problem in some types of structure is to produce the joint filler in situ and although this method can be used only when the joint can be formed as an air gap, some manufacturers are studying this approach.

Rigid and semi-rigid cellular boards can be produced from hard synthetic rubbers, in some cases by using a small amount of plasticizer to overcome brittleness, by vulcanizing a plastic or pliable material, or by adjusting the molecular weight of the polymer. The recovery of some of these materials may decrease under repeated or heavy loading and some may also tend to become distorted when stored in the open.

Considering the variations in properties which can be obtained with the synthetic resins already in general production it seems highly probable that a material combining all the desirable properties of joint filler and sealing compound will eventually be produced at an economic price.

Mineral fibre. Various mineral fibres such as glass fibre and slag wool have been formed into boards. The fibres are usually mixed with a suitable binding medium such as a resin or bitumen and are then enclosed between two layers of felt. When the amount of binding medium is the minimum necessary to fulfil its function of binding the fibres together, these fillers are compressible and do not extrude under compression. For some purposes they are perhaps compressed too easily, for example, under water pressure a surface sealing compound will be forced into the fibre core.

In common with other fibrous fillers they are permeable. They do not as a rule, however, swell during wetting.

Loose mineral fibre is of particular value as a packing material for maintaining expansion joints or for filling expansion gaps which have been formed during construction. It can be caulked sufficiently to provide support for a sealing compound whilst still being sufficiently cellular to accommodate expansion movement without extruding.

Metal. A method of filling expansion joints using thin gauge steel sheet which has probably been used only in pavings and bridge decks is shown in Fig. 6. 3. Whilst being rather expensive this method of filling joints might be appropriate to modern methods of road construction particularly if a semi-rigid plastic were used to form the expansion space instead of steel. The cap could also be designed to induce a crack above the centre of the joint when concrete

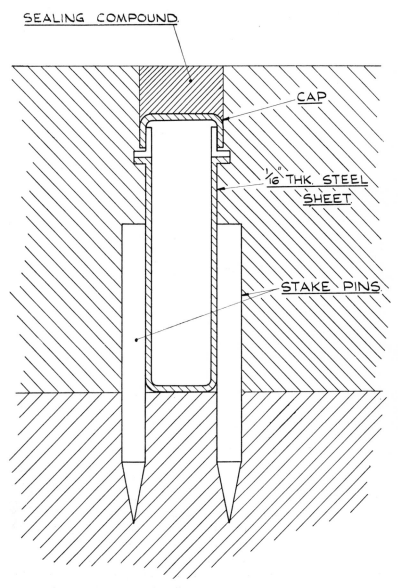

FIG. 6. 3.   A metallic joint filler which provides an air gap.[23]

is laid continuously and the sealing cavity is formed by saw cuts in the green or hardened concrete. In addition to defining the centre of the expansion space such a cap would also be more easily deformed and more resilient than a steel cap, thus preventing the development of dangerous stresses at the top of the joint when the concrete expands.

Choosing a joint filler

There is no material commercially available at the present time which completely fulfills all the requirements of a joint filler. It is desirable, therefore, to decide which properties are of particular importance in the type of structure in which a joint filler is to be installed.

(a) *Water-retaining and buried structures.* In these structures the joint filler must, above all, provide adequate support for the sealing compound in order to prevent it from being displaced by water pressure acting on the surface. Where this pressure is high it is desirable to use a relatively incompressible joint filler requiring a force in the region of, or perhaps slightly higher than, the tensile strength of the concrete to compress it to half its original thickness. In view of the massive or heavily reinforced construction used in water-retaining or buried structures the use of such a filler is unlikely to cause harmful stresses during the construction period when the structures are exposed to atmospheric temperature variations. Once in service the structures are protected from these variations and only small movements occur.

For ease of installation, the joint filler to be used in sea walls, venturi flumes and other structures having irregular cross-sections should be available in large sheets and should be easy to work.

Although the joint filler should not normally be in contact with the water contained by a structure it is often considered desirable that it be impermeable and that its compressibility be unaffected by contact with water. Similarly in structures carrying drinking water it may be required that the joint filler impart no taint to the water.

(b) *Buildings.* Due to the extra cost involved in erecting double columns and beams in order to provide expansion joints in buildings and to the difficulties sometimes experienced in concealing these joints, it is customary to economize on the number of joints to be provided. Being fully exposed, however, large movements can frequently occur, particularly at the top of the walls and in the roof of a large building. In order to prevent displacement of the sealing compound and disfigurement of the building, therefore, it is essential that the joint filler be easily compressible and non-extruding. The filler should also be sufficiently rigid to permit installation in vertical joints formed between split or double columns.

TABLE 6.1. RECOMMENDED EXPANSION JOINT FILLERS FOR VARIOUS CONCRETE STRUCTURES.

Type of Structure	Types of Joint Filler									
	Soft cellular plastic	Rigid cellular plastic	Cork strips	Granulated cork board	Dehydrated cork	Soft wood free from knots	Impregnated fibre board	Chip board	Cellular rubber	Solid rubber
Reservoirs and Dams				†		†	†	†		
Sea walls and other structures having irregular or curved cross sections				†		†	†	†		
Buildings	†	†	†	†	†	†	†			
Pavings	†	†		†	†	,	†			
Concrete pipes	†								†	†
Masonry cladding		†	†	†			†			
Bridge decking				†	†	†	†			

† Recommended materials.

69

When curing is effected by periodic hosing down with water it is necessary to ensure that this treatment does not result in the joint filler staining the concrete adjacent to it. No examples of staining due to this cause have been encountered in temperate climates but it is an effect which may take place where high temperatures are experienced.

The adhesive cellular plastic joint fillers which also provide a seal against penetration of water should, if perfected, be particularly suitable for use in expansion joints in the walls of buildings. It will, however, be necessary for these materials to be maintained in a state of compression or to be firmly bonded to the concrete to prevent displacement by gravity from joints in overhanging parts of a building, e.g. the underside of balconies. These materials are not likely to be suitable for forming expansion joints on account of their flexibility but they could be used in conjunction with a conventional joint filler. When an open expansion space is formed during construction these materials can be used to fill and seal the joint to a depth of about one inch.

The formation of an open space to accommodate expansion is often convenient in the construction of a building. The need for a joint filler is thus overcome and the space can be filled and sealed either by means of a suitable form of weather strip set into the adjacent sections during concreting or by a flexible capping strip inserted after the concrete has hardened and initial shrinkage has taken place. " Dry " joints used in curtain wall structures operate on these principles.

Pavings. When pavings are constructed continuously using heavy vibrating machinery it is essential either that the joint filler be rigid or that it be provided with adequate support during laying to prevent it becoming damaged or displaced as the machines pass over it. The filler should also be capable of compression to half its original thickness by a force in the region of the tensile strength of the concrete and should not extrude or release impregnant during such compression.

Recommendations for materials which may be considered suitable for use as joint fillers in various types of structure are given in Table 6. 1.

CHAPTER SEVEN

Jointing Materials — The Sealing Compound

Sealing compounds are usually thermoplastic materials and they can be used in almost every type of joint. They are often applied to a slot at the face of the joint but may in some methods of construction be poured into a cavity formed within the joint.

The functions of a sealing compound are as follows :—

(1) To seal the joint against the passage of water.

(2) To prevent the ingress of grit or other foreign matter.

(3) To provide protection to the joint filler where necessary.

In some types of joint, for example spigot and socket pipe joints, the sealing compound may be required also to provide a cushion between abutting sections. Most conventional compounds, however, are basically fluids and will flow when pressure is applied to them. It is desirable in such joints, therefore, either to reinforce the sealing compound with a material capable of withstanding pressure, or to employ a joint filler, such as rubber, which will also provide an effective seal.

THE PROPERTIES OF SEALING COMPOUNDS

The properties required in a sealing compound will vary according to the conditions under which it is to be used. The most important properties which apply in varying degrees under nearly all conditions are as follows :—

(i) *Impermeability*. As the main function of a sealing compound is to prevent water passing through the joint the material must be impermeable to water even when under considerable pressure. The culverts in power stations for example are frequently tested under a 100 ft. head of water.

It is the need for this property which results in fluids being most commonly chosen as sealing compounds although certain thermoplastic materials not generally considered as fluids are also impermeable.

(ii) *Adhesion*. In order to fulfil any of its functions a sealing

71

compound must develop an effective adhesive bond or contact under pressure to the walls of the joint to which it is applied. The adhesion of a sealing compound, whilst being partly dependent upon the viscosity of the material during application and upon its composition, is also considerably influenced by the nature and condition of the surface to which it is applied. A certain amount of mechanical bond may occur on rough surfaces but this will be effective only if the consistence of the sealing compound is such as to promote complete contact with the surface using a practicable method of application.

In general, the adhesion of a sealing compound will vary inversely with the viscosity of the binding medium and the proportion of inert material it contains. The adhesion of a sealing compound can frequently be improved by the application of a suitable primer.

(iii) *Deformability.* A sealing compound must be sufficiently deformable to accommodate the amount and rate of movement occurring between the sections of concrete on each side of the joint. This property is particularly important in joints which may be exposed to low temperatures or to high rates of strain as for example in exposed pavings carrying fast traffic.

It would appear that, in addition to low viscosity, elastic characteristics are desirable to enable a sealing compound to accommodate rapid movements at low temperatures.

(iv) *Resistance to flow.* Although it is necessary for a sealing compound to deform readily in order to accommodate the movements which may occur at the joint it is often important that the compound should not flow due to the effects of gravity.

Sealing compounds which are used in vertical or steeply inclined joints must have high resistance to flow. To obtain this property the binding material of the compound must have a high apparent or real viscosity particularly if the joint is likely to be exposed to high temperatures.

Obviously it is not possible to produce a compound which has a low viscosity at low temperatures and becomes more viscous as its temperature increases. The nearest approach to this ideal characteristic would be a compound whose consistence was unaffected by variations in temperature and it is along these lines that development work on sealing compounds has progressed.

(v) *Resistance to the ingress of foreign matter.* In many types of structure a sealing compound should not only resist the entry of water but also the ingress of grit, stones and other foreign matter. A similar effect which may be considered under this heading is indentation by the steel or cast iron wheels of trolleys, legs of stools, etc.

Sealing compounds composed of high viscosity fluids may be

resistant to the ingress of foreign matter but such compounds will not usually be sufficiently deformable to accommodate the movements of the joint at low temperatures.

The ideal properties in a sealing compound to resist these effects are a combination of elasticity and low temperature susceptibility.

(vi) *Resistance to weathering.* Sealing compounds which are exposed to the weather should be resistant to the oxidizing effect of the combination of sunlight and air. This is a surface effect, however, and as it is usually possible to provide a substantial depth of sealing compound in a joint, weathering of sealing compounds is rarely a serious problem except in those materials containing light oils.

(vii) *Chemical resistance.* Occasionally it is necessary for a sealing compound to be resistant to acids and organic solvents. In general, concrete itself is less resistant to attack by acids than conventional types of sealing compounds so that little attention is necessary to this property as far as sealing compounds are concerned. Resistance to the effects of oils and organic solvents, however, must be considered when choosing a sealing compound because many sealing compounds are based on oils and resins which are susceptible to attack from these materials particularly under prolonged exposure.

Many of the properties which would exist in a good general purpose sealing compound are incompatible. In most applications, however, greater emphasis is necessary on one or two of the general properties and a sealing compound can usually be found which possesses these properties to a high degree but lacks other less important properties.

TYPES OF SEALING COMPOUND

At the present time there are six different types of sealing compound available and these may be classified as follows :—

Hot-poured materials, which are usually bituminous compositions.

Bituminous compositions in the form of strips or pastes which are rendered workable by heating.

Cold applied resinous or bituminous compositions containing oils and volatile solvents which render them workable.

Bituminous emulsions to which setting agents are added immediately prior to application to the joints.

Prefabricated strips of resinous or bituminous compositions which develop adhesion when pressed against a clean dry surface.

Synthetic polymer materials which may contain oils or bituminous compositions and which set with the aid of heat or a chemical which accelerates the setting process after application to the joint.

Within each of these classes materials of widely differing composition and performance may be found. It is possible, however, to consider the basic characteristics of a number of typical sealing compounds in each class.

Hot-poured compounds

(i) *Residual bitumen.* These compounds are manufactured from various grades of bitumen obtained in the distillation of crude petroleum oil. They frequently contain non-volatile fluxing oils and mineral filler in order to reduce their tendency to flow in hot weather and to become brittle in cold weather. Fibrous fillers are particularly efficacious but they tend to impair application properties. Residual bitumen compounds are easy to pour and will adhere well to concrete, steel and other structural materials providing they are applied to clean dry surfaces. They may, however, tend to become displaced by water.

(ii) *Oxidized bitumen.* The properties of residual bitumen can be considerably modified by passing air or steam through the molten material while it is still in the retort or subsequently. In this process oxygen combines with the bitumen which then exhibits pronounced elastic properties and is less affected by changes in temperature.

Sealing compounds manufactured from oxidized bitumen in common with those based on residual bitumen may contain fluxing oil and filler. Full scale performance appears to indicate that, due possibly to greater internal cohesive strength, the adhesive properties of oxidized bitumen compounds are somewhat inferior to those of residual bitumen. They are usually easy to pour and their elastic properties cause them to resist to some extent the ingress of foreign matter and permanent deformation due to transient loading.

(iii) *Rubber/bitumen mixtures.* The greatest improvement in the properties of bitumen for use as a sealing compound is achieved by the incorporation of natural or synthetic rubber in the form of a fine powder or a latex. Most rubber bitumen sealing compounds are suspensions of swollen particles of rubber in a relatively soft grade of residual or oxidized bitumen. The effect of these particles is to impart elasticity to the compound and to reduce its temperature susceptibility. The decrease in temperature susceptibility is so marked in compounds having a high rubber content that difficulty may be experienced in rendering them sufficiently fluid for pouring into the joint without causing deterioration. Such deterioration, however, should become apparent only after the compound has been maintained at a temperature in excess of about 200°C. (390°F.) for a number of hours and under normal working conditions this rarely occurs.

A more common cause of deterioration due to overheating, which may occur with any hot-poured compound, is the application of fierce heat during the initial melting period when only a very small amount of material is in a molten condition. During this period circulation due to convection currents is negligible and furthermore it is usually difficult to stir the compound effectively. Considerations such as these have led to the development of oil-jacketed heating and pouring appliances which will be considered in detail in a later chapter.

The adhesion of rubber-bitumen compounds varies with the viscosity and type of bitumen used in their manufacture and the amount and type of rubber they contain. Providing the walls of the joint are properly prepared satisfactory adhesion can usually be obtained.

(iv) *Rubber/tar and rubber/resin mixtures.* The need for sealing compounds which are not readily attacked by a wide-cut paraffin aviation fuel has led during the last few years to the development of rubber/tar and rubber/resin mixtures. As natural rubber, (a) does not form stable mixtures either with coal tar or with most resins and (b) is soluble in oil and paraffin, synthetic rubber is used in the manufacture of these compounds.

Dispersions of synthetic rubber in coal tar are probably the most successful hot-poured sealing compounds available at the present time. In addition to their resistance to attack from oils and solvents these compounds are elastic ; they have low temperature susceptibilities and they adhere well to most clean dry surfaces. In order to obtain and preserve these properties, however, particular care must be taken in their manufacture. and in melting and pouring them on the site. The use of indirectly heated melting and pouring appliances is particularly recommended for the preparation and application of these compounds.

(v) *Resinous materials.* Natural and synthetic resins usually blended with fluxing oils and containing a high proportion of mineral filler have been used more particularly in an attempt to produce sealing compounds which will blend with their surroundings. The use of light-coloured ingredients permits almost any colour to be obtained in these compounds. Apart from their aesthetic appeal, however, no hot-poured resinous compound has been found to give a lasting good performance in service.

The chief faults in resin-based hot-poured sealing compounds are due to the properties of the resins themselves which usually have high temperature susceptibilities and are, in addition, frequently prone to weathering effects.

Light-coloured sealing compounds are, however, much to be desired and the rapid advances in the field of synthetic resins may

75

well result in the production of compounds which will be not only aesthetically satisfying but also efficient sealing compounds. Cold applied materials are already available which promise to satisfy all these requirements and it appears likely that these materials will eventually replace conventional hot-poured compounds.

Bituminous strips and pastes. These compounds have been developed for sealing vertical or suspended joints. Hot-poured compounds may not readily be applied to these types of joint and after application invariably tend to flow out of the joints if not restrained from doing so by means of a rigid cover.

Several bituminous pastes are available and these are mixtures of bitumen or coal tar blended with non-volatile oils and containing a high proportion of mineral filler which usually includes asbestos. The pastes are rendered workable either by direct heating or by immersion in hot water and are pressed into the joint by hand caulking tool or a trowel. The ease of working varies considerably.

Due to the high proportion of filler it must contain to prevent slump in vertical joints, a paste type sealing compound must have a low viscosity binding medium in order to promote adhesion.

The performance of bituminous pastes is generally satisfactory providing they are thoroughly worked into the joints during application.

Only one strip form sealing compound is known which is applied with the aid of heat. This compound is a rubber/bitumen composition containing asbestos and is supplied on reels, successive layers of the compound being separated by treated paper. The compound is applied to the joint by heating the surfaces which are to make contact with the base and the walls of the joint and working the material into the joint using heated irons. One disadvantage of a compound supplied in this form is that the joint widths are never absolutely uniform and this leads to difficulties in applying the compound satisfactorily. When properly applied this compound gives very satisfactory performance particularly at high ambient temperatures.

Resinous and bituminous compositions containing solvents. The largest group of materials of this type is the building mastics. These are mostly blends of resin or bitumen and various vegetable or mineral oils some of which are volatile and serve more particularly to render the materials workable. They usually contain a high proportion of fillers to retain the oils and to inhibit slump when in the joint. They may be applied by gun or by trowel and subsequent to their application evaporation of the volatile constituents or oxidation of oils results in the formation of a skin which provides a protective layer at the surface of the compound. Due to their methods of application and the choice of colours made possible by the use of light-coloured resins, these compounds are very suitable for use in buildings and the treatment of cracks.[24] Being of a soft consistence, adhesion to the walls of a joint or crack is usually good and the compounds deform readily in order to accommodate the movements of adjacent sections of a structure. They tend to harden in service due to loss and oxidation of the oils and resin, the rate of hardening depending on the choice and quantity of the oils they contain. Painting the surface with a good quality exterior oil paint considerably reduces deterioration.

Some building mastics tend to flow out of the joint either immediately after application or subsequently and to minimize this possibility they should be used only in joints $\frac{3}{8}''$-$\frac{1}{2}''$ wide unless some form of reinforcement is set into the material during application.

Building mastics may be supplied in tins or drums or in small cellophane or cardboard cartridges for insertion into a gun which extrudes the compound through a nozzle by means of a trigger-operated piston.

Another type of material[25] in this class is supplied in two parts which are mixed together on the site immediately prior to application. One of these parts comprises a blend of residual bitumen, non-volatile and volatile oils and the other is a mixture of hard bitumen in powder form, mineral filler and possibly powdered rubber. This type of material has good adhesive properties but the proportion of fluid to powdered ingredients must be carefully chosen in order to avoid slump immediately after application or the development of brittleness and excessive shrinkage as the volatile oils evaporate.

All sealing compounds which rely on solvents to render them workable shrink as the solvent evaporates. The volume of compound used in a joint should be restricted therefore to the minimum necessary to accommodate the movements at the joint without failure in adhesion. A cross-section of $\frac{1}{2}''$ x $\frac{1}{2}''$ is satisfactory for most materials having a soft binding medium.

Emulsions. The difficulties experienced in pouring hot-applied sealing compounds having low temperature susceptibilities led to the production of these compounds in the form of emulsions.

Bituminous emulsions, often containing rubber in the form of latex, have been used with varying success as joint-sealing compounds.

The method of using an emulsion is to mix it immediately prior to application either with a powder which reacts with the water in the emulsion or with an acid which will cause the emulsion to coagulate. Compounds of very low temperature susceptibility can be produced by either of these methods.

The stability of an emulsion varies according to atmospheric temperature and moisture conditions. Close control is therefore

77

necessary in the manufacture of emulsion type compounds and it is difficult to assess the amount of setting agent necessary under a given set of conditions to produce the rate of coagulation appropriate to the method of application. The use under warm dry summer conditions of the same amount of setting agent as that used under cold or humid conditions will result in rapid coagulation and it may be found impossible to apply the compound to the joint. Conversely if insufficient setting agent is added under cold or damp conditions the compound will be too fluid and will be impossible to apply, for example, to vertical joints.

In the hands of a skilled operator, however, excellent results may be obtained with rubber/bitumen emulsions and moreover as they contain water it is not essential for the surfaces to which they are applied to be absolutely dry. Probably the most common fault in these compounds is the tendency for them to slump in vertical joints if they have not been correctly apportioned or mixed prior to application.

Adhesive pre-formed strips. In addition to the foamed plastic materials already described in Chapter VI under the heading of joint fillers there are four types of pre-formed strips which are intended specifically as sealing compounds. All these compounds rely, in varying degrees, to pressure being exerted on them by the adjacent walls of the joint.

The types of strip are as follows:—

(i) *Treated fabrics.* These are strips of fabric which are thickly coated or impregnated with a soft mineral grease or which are employed to contain a bituminous mastic. They may be used when a joint width of up to ⅛ inch is required and being soft they develop good adhesion on contact so that pressure is required only to take up irregularities in the walls of the joint. These strips contain a small amount of deformable material and are therefore suitable only in joints accommodating small movements.

(ii) *Rubber/bitumen compounds.* Both natural and synthetic rubber may be masticated in bitumen in such a way as to produce compounds which may be extruded into strips. Being dimensionally stable, strips having various cross sections are available. Preformed rubber/bitumen strips have marked elastic properties and high viscosity. When adequate pressure is available to deform the compound a very strong adhesive bond may be obtained to most surfaces which are clean and dry. These characteristics make this type of preformed strip applicable to joints of almost any width and thus a considerable range of movement may be accommodated.

(iii) *Natural and synthetic resins.* A wide range of consistence may be obtained in compounds manufactured from natural and synthetic resins. Strip form sealing compounds have therefore been produced using natural resin, oil and appropriate fillers, a blend of natural and synthetic resin, oil and fillers and plasticized synthetic resins and fillers, in particular, using butyl rubber.

The deformability of these three types of resinous sealing compounds varies considerably according to the quantity and type of fillers and fluxing or plasticizing oils they contain. Usually, however, they are intermediate between the treated fabrics which are very soft and the rubber/bitumen strips which are relatively hard and elastic. The compounds containing synthetic resins may also have elastic properties.

These strip type compounds are suitable for sealing joints where only a limited amount of pressure may be brought to bear on the compounds and find a particular application where large irregularities must be accommodated when the joint is made. The amount of movement they will accommodate will vary considerably depending on their adhesive properties and consistence. In general their adhesive properties are not as good as the other types of strip sealing compound and therefore they will accommodate only very limited joint movements. Some of the synthetic resins tend to become brittle on exposure to sunlight and air.

(iv) *Preformed natural and synthetic rubber.* Typical cross-sections of preformed sealing strips made from natural and synthetic rubber are shown in Fig. 7. 1. Solid strips, usually of circular cross-section, are also used but more particularly in pipe jointing. These strips have no adhesive properties and they must therefore be maintained in a state of compression at all times in order to prevent water from entering the joint. The channel section has been used in buildings only and whereas it may provide a first barrier to water infiltration a more conventional sealing compound should preferably be provided underneath it.

The hollow strips have been used to seal joints in pavings as well as buildings and they have the advantage that they may be evacuated before being inserted in the joints. Whereas, however, they possess most of the properties required for sealing joints in pavings it is essential that the joints be of a

NATURAL RUBBER.

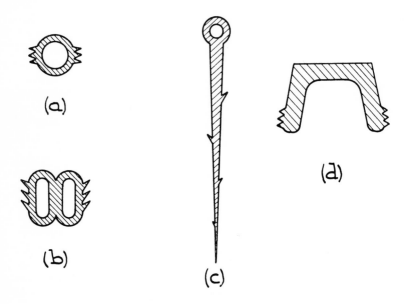

SYNTHETIC RUBBER

FIG. 7. 1. Sealing strips.

uniform width in order to avoid displacement due to differential vertical movement between adjacent slabs. Observations made on a number of full-scale experiments to assess the performance of hollow rubber strips have indicated[26] that, in spite of variations made in the shape of the sealing cavity to prevent displacement, none of the strips could be relied upon to prevent water from entering the joint.

A particular class of preformed synthetic rubber strip which can also have adhesive properties is an impregnated foam such as bitumen or butyl rubber impregnated polyurethane.

This type of strip has inter-connecting cells and according to the amount of impregnant and the degree of compression it will withstand quite substantial heads of water provided that it has adequate backing support. A typical design graph for such a material is given in Fig. 7. 1.

Setting synthetic resins (*elastomers*). Hitherto synthetic resins have been used principally for the manufacture of a variety of moulded and extruded articles or the production of material in simple forms for subsequent working into a finished article. The production of the resin in its final chemical state has always been carried out in a factory. Recently however, it has been found that some resins can be supplied with advantage in the form of pastes which are set in situ by heat treatment " plastisols " or by chemical reactions (*Thiokols*, polyurethanes, epoxies and silicones).

(a) *Plastisols.* There are not likely to be many occasions in civil engineering or building construction when heat can be applied with any degree of control to a sealing compound in a joint. It is possible, however, that such opportunities may occur in prefabricated work and resinous pastes requiring heat treatment should therefore be included as possible sealing compounds. Their chief applications, however, are in coach and motor body building and in the treatment of fabrics. These pastes are commonly known as " plastisols " and are mixtures of synthetic resins (vinyl polymers) in the form of regular grain size powders, plasticizing oils and usually inert fillers.

The method of using this type of compound is to maintain it at a temperature of about 150°C. (300°F.) for a period of 5-20 minutes immediately after application. The resin and plasticizing oil then combine and the compound sets to an elastic solid.

Plastisols adhere well to metals and it is possible that they could be used to form a prefabricated joint between metal sections which are to be set in concrete. Due to the high temperature curing process involved, these compounds have not been used in conventional concrete joints and it is unlikely that plastisols as used at the moment merit serious consideration for such applications.

81

F

(b) " *Thiokols.*" These are polysulphide polymers in the form of a stiff paste or a liquid which when mixed with a catalyst, set to an elastic solid. When used as a sealing compound they may be mixed with another synthetic or natural material such as epoxy resin, tar or bitumen in order to impart adhesive properties. Additional filler may be incorporated as an extender.

In practice, the ingredients may be supplied in separate parts, a base and hardener, which are mixed on site; they may be premixed and supplied in a frozen state to inhibit the setting reaction or they may be in the form of an inhibited single pack material.

Usually these compounds are gunned into the joint but pouring grades, particularly when modified with coal tar, are also available.

The period during which a two-pack Thiokol compound remains workable after mixing varies between 20 minutes and several hours depending on its composition. They have excellent adhesive properties to most clean dry surfaces but adhesion falls off with time from mixing.

The temperature accommodation of these compounds can be very great, a typical range being −40 to + 160°F. (− 40 to + 70°C.). As these compounds usually contain a high proportion of synthetic rubber they have very pronounced elastic properties. The synthetic rubber however, is rather costly at the present time and in order to reduce costs high filler contents may be necessary which will result in a reduction in elasticity. This reduction in elasticity, however, may not have a deleterious effect on the performance of the compound providing that it is not accompanied with a reduction in adhesive properties.

Insufficient experience has been gained with these compounds to indicate their resistance to natural weathering but laboratory investigations show them to be immune from attack by most agencies including oils, many of the common solvents, acids and alkalis. It would therefore appear that these compounds will remain effective for a long period of time even if exposed to very adverse conditions and in spite of their high initial cost they may be more economical than the cheaper but more conventional compounds which often require periodic replacement.

One notable exception is where the Thiokol abuts on to glass through which sunlight is passing to strike the glass/Thiokol interface. Under these circumstances the adhesion of the sealing compound is impaired by ultra-violet radiation. This effect can be overcome by the use of special primers which " block " the ultra-violet rays at the glass face.

(c) *Other elastomeric compounds.* Following the development of the " Thiokols " a whole range of elastomeric sealing compounds is continuing to develop. This range includes the following:—

Butyl and polyisobutylene (single pack)
Polyurethane (single and two pack)
Silicone (single pack)
Chlorinated polyethylene (single pack)
Flexible epoxy (two pack)
Chloro-butyl (two pack)
Ethylene-propylene.

Many compounds based on these polymers are under development and at this stage it would be misleading to detail the published properties of all of these materials because in many cases their compositions have not been finalized and practical applications have not been confirmed. Typical characteristics of some of the more firmly established products are given in Table 7. 1.

CHOOSING A JOINT SEALING COMPOUND

In the interests of economy, both in initial cost, and subsequent labour and material costs which may arise from the need for maintaining joints, it is essential that very careful consideration be given to the choice of sealing compounds.

The properties required in a sealing compound vary according to the type of joint and structure in which it is to be used, and whereas many joints in structures must be given individual attention there is a number of general principles to assist in the choice of appropriate sealing compounds.

Water-retaining structures. The principal function of the sealing compound in joints in a water-retaining structure is to prevent water passing through the joint. The sealing compound used in these structures should develop a very good adhesive bond to the walls of the joint. When considering bituminous materials, therefore, those having a low viscosity binding medium will be of particular interest. This does not mean, however, that compounds of the building mastic type are suitable. These compounds may contain volatile solvents which will contaminate drinking water and furthermore on evaporation of these solvents serious shrinkage effects will occur in the large sealing cavities which must be used in these structures to ensure that the joint is sealed against water under pressure. They may also fail in adhesion due to saponification of the oils they contain.

The movement to be accommodated by joints in water-retaining structures is generally small so that although materials chosen for their adhesive properties may also be readily extensible this latter property is of no consequence except that it reduces the bond stress developing as a result of any movement which does occur.

Resistance to flow is of only limited importance except during the construction period when the sealing compound may be exposed to the sun. It is important, however, to bear in mind that sealing compounds are generally fluids and will flow when anisotropic pressure is applied to them. When exposed to water pressure on one face a sealing compound must receive adequate support from the walls and the base of the sealing cavity. The sealing compound will offer negligible resistance to water pressure from within the joint and if such a condition can be anticipated a waterbar should be installed across the joint during construction.

Many hot-applied bitumen sealing compounds may be expected to give a reasonable performance in horizontal joints in water-retaining structures ; unfilled rubber/bitumen compositions being among the most effective. In vertical joints bitumen and rubber/bitumen compounds applied hot either as a preformed strip or in a putty-like consistence are satisfactory. Compounds which can be rendered sufficiently fluid by heating to enable them to be poured, tend to flow out of vertical joints and therefore are not satisfactory. Many of the elastomeric compounds are also suitable but their cost may not be warranted.

In open culverts and sea walls a sealing compound is often exposed to more severe conditions than in reservoirs and sewage treatment tanks where the compound is continuously protected by water or sewage. The conditions in sea walls are particularly severe because not only must the sealing compound resist abrasion but also it must be readily extensible to accommodate the large and sudden movements which occur when the walls are washed by the sea at high tide. The prying fingers of children and even adults must also be considered as possible destructive forces ! In the light of past experience it would appear that none of the conventional sealing compounds, used singly, can withstand these conditions and frequent renewal is necessary. Two promising methods have recently been evolved ; one of these is an elastomeric compound applied cold and the other is the hard rubber/bitumen strip which may be used to protect a soft bitumen/ asbestos sealing compound applied as a hot putty to provide an effective adhesive for retaining the protective strips as well as a readily extensible seal.

Buildings. The properties required in a sealing compound for use in a building vary considerably according to the part of the building in which the compound is to be used and it is convenient therefore to consider the requirements of the joints in walls, roofs and floors separately.

(i) *Walls.* One of the main considerations when choosing a sealing compound for the walls of a building is the appearance of the finished seal and for this reason resin based building mastics are often used. The amount of movement occurring at the joints in the walls of a building is often quite small so that narrow joints can be adopted and shrinkage due to loss of solvents is negligible with a good quality mastic. The particular properties of building mastics which make them suitable for sealing joints in the walls of buildings are as follows :—

(a) They are light in colour and may be pigmented.

(b) They can be applied cold by means of an extruding gun, thus enabling a neat finish to be obtained.

(c) They may be painted once a surface skin has formed.

(d) They are sufficiently extensible to accommodate joint openings up to 20% which is usually adequate.

(e) They will develop a good adhesive bond with most clean dry surfaces.

Although a building mastic may have very good properties initially and may also give a satisfactory practical performance for periods up to possibly 20 years, this performance must be considered in relation to the life of the structure and the cost of replacing the material when it has become unserviceable.

Weathering effects cause building mastics to become less deformable so that eventually movement at the joint will cause them to fracture or fail in adhesion. Another result of weathering, which takes place over a shorter period, is a substantial darkening in colour.

The cost of replacing a sealing compound in a large building can be very high and it seems probable therefore that the recently developed elastomeric compounds may be particularly appropriate. These compounds whilst possessing most of the desirable features of conventional building mastics are more promising from the point of view of longer satisfactory performance.

Adhesive cellular strips and preformed synthetic rubber strips of the type shown in Fig. 7. 1. are also worthy of consideration, particularly for the treatment of the joints inside a building where it is necessary to provide only a neat appearance. A further approach to the sealing of joints in the walls of buildings is the use of open drained joints. This method will be considered in more detail in relation to curtain walling.

(ii) *Roofs.* Being fully exposed large movements due to temperature variations can occur in flat concrete roofs particularly when a black surfacing has been applied for waterproofing purposes. It is essential therefore that adequate provision be made for this movement and that the joints be efficiently sealed.

In addition to the movement which must be accommodated it is necessary for the joint sealing compound to be resistant to flow into the joint when the concrete contracts in cold weather and in some structures it must also be capable of carrying foot traffic.

Sealing compounds for use in the joints in roofs must therefore have low temperature susceptibilities and good adhesive properties. Polysulphide pitch, rubber/bitumen and bitumen/asbestos compositions designed for use in vertical joints are among the most satisfactory materials.

(iii) *Floors*. The main function of the sealing compound in the joints in the floor of a building is often to protect the edges of the adjacent concrete slabs. In some buildings it may be necessary for the compound to be resistant to oils and fats. Adhesive properties are often of minor importance because the movement at the joints is usually small due to their close spacing and it is not usually necessary to prevent water from entering the joint. Under these circumstances a relatively hard strip of synthetic rubber can be employed. If some resistance to the entry of water is required a relatively hard hot or cold poured synthetic rubber/tar composition may be used.

Pavings. The types of paving which influence the choice of a sealing compound according to the conditions imposed on it may conveniently be considered under the following headings :—

(i) *Roads*. There is no material available at the present time which will provide an effective seal to the joints in a road for a period of more than a few years.

The five main properties required in a sealing compound for satisfactory performance are as follows :—

(a) Good adhesion to the concrete so as to prevent the entry of water which may soften or wash away the subgrade.

(b) Good extensibility to accommodate the large movements due to expansion and contraction of long slabs and due to vertical deflections under the transient traffic loads.

(c) Good resistance to the ingress of grit, which if forced into the joints by traffic will cause spalling of the edges of adjacent concrete slabs.

(d) Resistance to flow so that the compound does not tend to flow out of joints on a steep transverse or longitudinal gradient or flow down into the joint when the joint filler fails to fill the joint completely.

(e) Ease of application so that the compound may be applied to fill sealing cavities as narrow as 1/8-3/16 inch.

Records of full-scale performance indicate that hot and cold applied rubber/bitumen and rubber/tar compounds give the best all round performance particularly if suitable mechanical aids are used to apply them. It is likely, however, that elastomeric materials may be more applicable to the narrow sealing cavities formed by sawing joints in the hardened concrete.

Very severe conditions exist on housing estate roads during house construction. During this period the roads are covered with ballast and grit which is pounded into the sealing compound by heavy construction vehicles. It is advisable on such sites either to set a protective fillet of wood or hard rubber/bitumen strip into the sealing compound when it is laid[27] or to fill the joints to the surface with the joint filler and apply the sealing compound after the construction traffic has left the site.

(ii) *Airfield runways and other paved areas.* Although vertical deflections and horizontal movements may be larger at the joints in runways and similar paved areas carrying high wheel loadings, the properties required in sealing compounds for these joints are to some extent less conflicting than those for road joint sealing compounds. These pavings are normally laid to falls which are only just sufficient for drainage purposes so that there is no tendency for the sealing compound to flow out of the joints. It is customary also to sweep from the surface of these pavings all loose grit which would not only be a hazard to flying operations, but would also penetrate into the sealing compound and cause spalling of the concrete. This relaxation in the conditions permits the use of softer materials which will accommodate movement more readily and will also have good adhesive properties. Rubber/bitumen compounds having a relatively low viscosity are therefore generally satisfactory.

At the ends of airfield runways and on hard-standings jet aircraft create another set of conditions. In these areas it is desirable that a sealing compound be resistant to the wide cut paraffin fuel which is released from the jet engines when they are shut down and which may also be spilled during refuelling operations. It is also desirable that the compound have some resistance to the heat and blast of jet engine exhausts.

Resistance to fuel and oils is also necessary in compounds used to seal the joints in concrete garage forecourts and car parks.

Synthetic rubber/tar compounds are commonly used where these

TABLE 7. 1. THE PROPERTIES REQUIRED IN SEALING COMPOUNDS FOR VARIOUS TYPES OF STRUCTURE.

Type of Construction		Properties Required for Satisfactory Performance					
		Adhesion	Extensibility	Resistance to Flow	Resistance to Ingress of Foreign Matter	Resistance to Weathering	Resistance to Oil, Fuel and Fats
Water Retaining Structures	Vertical Joints	Very Good	Fair	Very Good	Not Important	Not Important	Not Important
	Horizontal Joints	Very Good	Fair	Not Important	Not Important	Not Important	Not Important
	Open Culverts	Very Good	Fair or Good	Very Good	Good or Very Good	Good	Not Important
	Sea Walls	Very Good	Very Good	Very Good	Very Good	Very Good	Not Important
Garage and Factory Floors		Sometimes necessary	Fair	Not Important	Very Good	Not Important	Sometimes Essential
Buildings	External Joints	Good	Fair or Good	Very Good	Fair	Very Good	Not Important
	Internal Joints	Sometimes necessary	Fair	Very Good	Good	Not Important	Sometimes necessary
Pavings	Trunk Roads	Very Good	Very Good	Fair	Good	Very Good	Not Important
	Housing Estate Roads	Very Good	Good	Fair	Very Good During House Construction	Very Good	Not Important
	Runways	Very Good	Very Good	Not Important	Fair	Very Good	Necessary where fuel spillage occurs

TABLE 7. 2 TYPICAL PROPERTIES OF ELASTOMERIC SEALING COMPOUNDS.

Type	Nature	Tensile Strength (b.s.i.)	Elongation (%)	Modulus of Elasticity (b.s.i.)	Hardness (Store A)	Application	Movement Accommodation % Joint Width	Comments
Butyl Rubber 1 Component	Plastic & Elastic	115–250	50–250	—	25–50	Gun, knife, trowel or tape	50–100	Retain softness well. Adhesion not always adequate
Chlorosulphonated Polyethylene 1 Component	Elastic	100–175	50–100	—	15–25	Gun applied	50–100	Good weathering. Poor storage stability. Good adhesion
Flexible Epoxy 2 Component	Elastic	100–6,000	4–20	—	—	Gun or poured	0–10	High temperature susceptibility limits extensibility. Good adhesion
"Neoprene" 2 Component	Elastic				0–10	Gun applied	50–100	Limited experience. Low curing rate
Polysulphide 1 & 2 Component	Elastic	45–100	100–250	25–60	20–65	Gun or poured	50–100	Not tolerant of grease. Good durability. Adhesion assisted by primer
Polysulphide/ Pitch	Elastic & Plastic	—	50–100	—	—	Gun or poured	Up to 50	Non-sag or self-levelling grades available. Good weathering, temperature and solvent resistance
Polyurethane 1 Component	Elastic	70–110	130–250	25–75	30–50	Gun applied	50–100	Moisture sensitive in uncured state. Good weathering and solvent resistance. Good low temperature performance
Silicone 1 Component	Elastic	100–140	75–200	60–140	20–35	Gun applied	100–150	Limited experience. Moisture sensitive in uncured state. Good weathering. Large temperature range

89

conditions are met but there is scope for improvement in these compounds in order to increase their resistance to jet engine exhausts. Boldly applied elastomeric compounds show promise for these conditions.

SUMMARY OF THE PROPERTIES REQUIRED IN SEALING COMPOUNDS

A general indication of the more important properties required in sealing compounds for various types of structure is given in Table 7. 2. It is suggested that this table be used as a general guide only, permitting an initial selection to be made of the type of material which might be suitable for sealing the joints in any given structure. The compounds should be chosen finally in relation to the particular project under consideration.

CHAPTER EIGHT

Jointing Materials — The Waterbar

It is important in the design of many types of structure that every precaution be taken to prevent water passing through the walls, floor or roof. Appropriate design and construction methods must therefore be adopted either to prevent the development of cracks, or to ensure that any cracks which may occur do so at points where an effective seal has been provided during construction or can be installed subsequently.

Thus, although only a small amount of movement may be expected to occur in water-retaining structures such as covered reservoirs, underground culverts or tunnels and other protected structures, a sealing medium is often provided at construction joints and other possible sources of weakness.

In addition to the joints which may be necessary to accommodate shrinkage and thermal movements during the construction of these types of structure it might also be desirable to provide joints to accommodate movements due to unstable soil conditions or the loading conditions imposed by the overburden.

It may be possible to seal these joints with a suitable sealing compound applied to a slot formed at the surface but where such a compound is subjected to water pressure or to high rates of flow, small weaknesses in the seal may result in progressive deterioration and serious leakage. These weaknesses may be due to poor workmanship in preparing and filling the sealing slot or to porous concrete around the slot, where, for example, difficulties may have been experienced in placing and compacting the concrete.

Another method of sealing joints against the passage of water is to install a strip or plug of impervious material across the joint either within the section of concrete or at the surface. This type of seal is called a " waterbar " and may be used alone or in addition to a surface seal.

91

THE USES OF A WATERBAR

A waterbar should be provided across joints or planes of weakness in all structures under the following circumstances :—

(1) Where external ground water pressure may develop.

(2) Where water is impounded at a head of ten feet or over.

(3) Where it is essential that there shall be no risk of the penetration of water.

(4) Where formwork or the method of construction makes accurate formation of a surface sealing cavity difficult.

Waterbars which are cast into the concrete, however, should not be used in thin pavings designed to carry traffic because they form a continuous plane of weakness at the edges of adjacent slabs and this may lead to the development of cracks as illustrated in Fig. 8. 1.

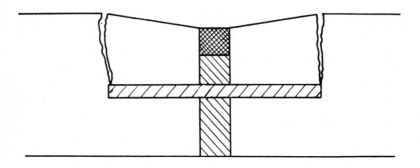

FIG. 8. 1. The type of failure induced by a waterbar in a paving carrying traffic.

THE FUNCTION OF A WATERBAR.

The only function of a waterbar is to prevent the passage of water through a joint or crack but it must at the same time permit movement not only at right angles to the plane of the joint or crack but possibly along this plane.

A waterbar may prevent water flowing through a joint or crack in one of three ways :

(1) It may adhere to the concrete on each side of the joint or crack.

(2) It may act as a valve, water pressure and movement of the concrete causing high stress concentrations between the waterbar and the concrete at one or more points on each side of the joint.

(3) It may provide a long and tortuous path for the flow of water.

Most waterbars are designed to act as valves. Their profiles are irregular so that they are anchored in the adjacent sections of concrete. Water pressure and movements of the concrete due to shrinkage or changes in temperature or moisture content cause high stress concentrations at the points of anchorage. This is shown diagrammatically in Fig. 8. 2.

It is likely that due to the protection provided by the water, and the backfilling and overburden where these exist, little if any expansion movement occurs in the concrete in a water-retaining structure when in service. A valve type of waterbar therefore is maintained in the stressed condition caused by shrinkage of the concrete during setting. Water pressure also may produce stress in the waterbar and this pressure itself will assist a well-designed waterbar in providing a seal.

The properties of a waterbar

The properties required in a material for use as a waterbar vary considerably according to the method by which the waterbar functions. In addition to the essential property of impermeability therefore there are a number of other properties which may be required in materials to enable them to fulfil the functions of a waterbar.

Deformability. A waterbar must deform at a stress which is substantially lower than the tensile strength of the concrete. A waterbar which acts as a valve must not become permanently deformed by the hydrostatic pressure or by the stress developed due to movement of the concrete.

Materials such as bitumen which are melted and poured into the joint to form a waterbar in situ are not capable of sustaining stress and the joint to which they are applied must be designed in such a way that continuous deformation due to water pressure cannot occur.

Adhesion to concrete. This is a property which is desirable in all types of waterbar but it is essential only in materials which exhibit permanent deformation under loading. As these materials are normally poured into a cavity which is formed in the concrete during construction, they should be capable of developing adhesion to concrete having a high moisture content.

Durability. Waterbars which are cast into a concrete structure cannot be replaced once the structure is in service. Although it may be possible to repair any leaks which may occur due to a waterbar becoming defective, therefore, it is highly desirable that the waterbar be as durable as the structure itself. In most structures the waterbar is operating under almost perfect conditions being fully protected from

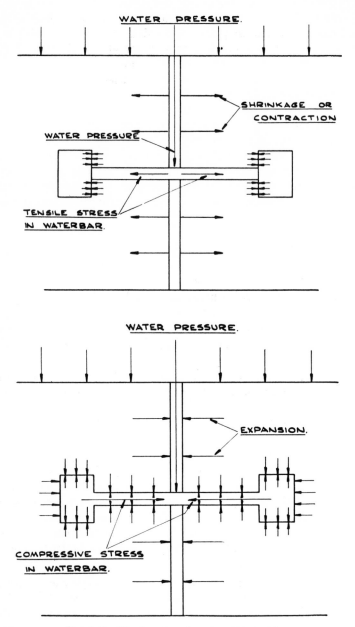

FIG. 8. 2. The function of a valve-type waterbar.

air and sunlight and maintained at an almost constant low temperature. Most materials deteriorate more rapidly when maintained in a stressed condition. The magnitude of the stress in a waterbar however is very low except in the few cases when severe movement must be accommodated and deterioration due to this effect is probably not significant. It is important however that continual flexing and conditions of continuous stress should not result in fatigue or work-hardening of the waterbar.

Ease of installation. The installation of a waterbar should not involve the use of complicated formwork and once set in position it should be sufficiently rigid not to sag under its own weight. A waterbar which is too flexible tends to become displaced by the concrete during placing and compaction ; such displacement may, in extreme cases, lead to structural failure of the concrete as well as serious leakage.

In many structures a waterbar may be used in joints which intersect, for example in lift joints and vertical construction or functional joints in walls. In order to provide an efficient seal the waterbar in such joints must be jointed so as to provide a continuous diaphragm ; the provision of a loose overlap is not likely to provide an efficient seal particularly if the joint is exposed to water pressure. It is essential therefore that watertight joints can be formed between sections of the waterbar either before or during installation.

Resistance to weathering. Although a waterbar is usually protected from weathering effects when in service it may occasionally be exposed to the weather for a relatively long period while a structure is being built. It should, under these circumstances, either be suitably protected, or have some resistance to weathering effects.

During the construction period the waterbar may also be exposed to other damaging agencies such as mobile construction plant. It is an advantage, therefore, if the waterbar can be conveniently stowed but failing this it must not be susceptible to mechanical damage.

Chemical resistance. A waterbar is not likely to be in contact with high concentrations of chemically active materials but in addition to the possibility of corrosion, which must be considered when metallic waterbars are used in water-retaining structures, there are other effects which must influence the choice of a water-bar for use in structures such as fuel oil bunkers and acid pickling tanks. These are special cases requiring individual attention however and the general requirements are that the waterbar should be resistant to chlorinated water, water having a high lime content and low concentrations of alkaline and acidic trade wastes.

TYPES OF WATERBAR.

Although there is only a limited number of materials which could be expected to fulfill the functions of a waterbar satisfactorily, many different profile designs are possible with each material. The best approach to waterstop design, therefore, is to consider the properties of available materials and then to examine typical designs which have been used in order to derive the greatest benefit from these properties.

Metallic Waterbars. The metals which have been most commonly used as waterbars are as follows :—

(i) Lead.

(ii) Ferrous metals.

(iii) Copper.

As these metals have significantly different properties their efficacy as waterbars must be considered individually. Typical designs of waterbar using these metals are shown in Fig. 8. 3.

(i) *Lead.* Lead waterbars are sufficiently flexible to accommodate the slow and usually relatively small movements due to changes in temperature and moisture content. They will not, however, accommodate differential transverse movement between abutting sections of concrete unless some degree of freedom is provided between the sections. Movement of the concrete does not result in a lead waterbar being maintained in a stressed condition and therefore lead is not a suitable material for use as a valve type of waterbar.

Lead waterbars deform very readily when a load is applied to them and when embedded in concrete they may be expected to accommodate movement without unduly disturbing the degree of contact, which might almost be considered a mechanical bond, between the metal and the concrete. When used in this way these waterbars thus restrict the flow of water through the joint by providing a long and very fine path at some distance below the surface of the concrete. Continuous contact with water containing lime results in the formation of a crust of insoluble salts on the surface of lead and it is often held that corrosion of this type assists in ensuring that lead and ferrous metal waterbars will prevent the flow of water through a joint or crack. It is highly probable however that movement occurring at the joint will fracture the layer of corrosion rendering it permeable.

A lead waterbar may also be laid on the surface of the concrete so as to span the joint, rag bolts being used to fix it firmly to the concrete on each side.

The chief disadvantage of lead as a waterbar is probably its complete lack of elasticity. Every cycle of flexing or extension and compression results in cumulative deformation which must eventually lead

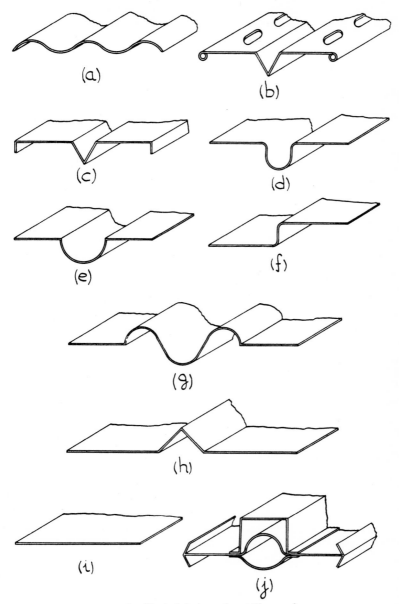

Fig. 8. 3.   Typical designs of metallic waterbars.

fracture. The useful life of a lead waterbar is thus governed completely by the amount and frequency of the movements occurring at the joint or crack.

Lead waterbars are liable to become damaged during the construction period.

(ii) *Ferrous metals.* Iron and steel strips often corrugated or formed into an irregular profile are sometimes used as waterbars, more particularly in construction joints where no movement is anticipated.

The elastic properties of steel render it suitable for use as a valve type of waterbar and providing that it is securely anchored in the concrete on each side of the joint very small movements of the concrete will provide high stress concentrations at the anchorage points thus resisting the flow of water. It is customary to coat these waterbars with bitumen in order to prevent corrosion.

The jointing of ferrous metal waterbars presents a difficult problem the only satisfactory solution to which is ribbon welding.

(iii) *Copper.* This is undoubtedly the most commonly used metal for the manufacture of waterbars. Indeed, copper waterbars have been used so extensively in structures all over the world, that for many years now they have been considered a convention. It would appear, therefore, from its continued use, that copper is probably the most satisfactory metal for this particular application. It is, nevertheless, very important that the properties of the metal be examined closely to determine its suitability for the function it is to perform and that the profile of the waterbar utilizes these properties in the best possible manner.

The most likely reason for the popularity of copper as a waterbar material is its known durability and resistance to corrosion. In addition to these properties however a waterbar material must often be capable of enduring almost continuous flexing, extension and compression effects due to the movements occurring at the joints in a concrete structure. The ability of copper to accommodate these movements over an extended period is limited by its work-hardening characteristics and it has been stated[28] that continual flexing of a copper waterbar will lead to hardening and crystallization which may result in fracture. The importance of these effects can be mitigated by suitable choice of waterbar and joint design and by appropriate treatment of the metal during manufacture.

Useful guidance on these aspects is given in British Standard Specification 1878 : 1952 " Corrugated copper jointing strip ". In addition to giving a profile ((d) in Fig. 8. 3.) which minimizes the possibility of stress concentrations in the copper due to movement, this Specification describes the type and gauge of sheet which should be used as follows :—

" *Quality of material.* The corrugated jointing strip shall be formed from half-hard copper strip complying with the requirements of B.S. 899, ' Cold rolled copper sheets and strip (half-hard and annealed) for general purposes.' The strip shall be manufactured from copper complying with the requirements of one of the following British Standards :—

" B.S.1172. Phosphorus deoxidized non-arsenical copper for general purposes.

" B.S.1174. Phosphorus deoxidized arsenical copper.

" If, however, the joints are not to be welded the strip also may be manufactured from copper complying with the requirements of either B.S.1040, ' Tough pitch copper,' or B.S.1173, ' Tough pitch arsenical copper '."

Depth of corrugation (in.)	Width of strip (in.)	Thickness		Depth of corrugation (in.)
		(in.)	S.W.G.	
1¼	8	0.022	24	1¼, 1½, 1¾, 2.
1½	10	0.028	22	1¾, 2
1¾	10	0.032	21	2.
2	10			

One disadvantage of crimped copper waterbars is that they cannot readily be bent in order to follow vertical or horizontal curves and in common with other metallic waterbars site fabrication of joints where intersections or changes in direction occur is a tedious operation.

Poured waterbars. Cavities are sometimes provided in the centre of joints for subsequent filling with a poured material such as bitumen, which then forms a waterbar. The properties of this filling material must be such that it may be rendered sufficiently fluid to enable it to be poured to fill completely a cavity possibly four inches in diameter and up to twenty feet or more deep. Being incapable of sustaining stress so as to form a valve type of waterbar the filling material must develop a good adhesive bond to the concrete. It should not contract on setting or tend to flow out of the joint or become brittle. It is impossible at the present time to obtain a material having all these properties and in practice the most commonly used material is bitumen, appropriate precautions being taken in the design of the joint to minimize its shortcomings.

The synthetic resin materials which can be polymerized in the joint, such as Thiokols, could probably be used as poured waterbars.

These compounds can sustain stress and thus could be poured to form a valve type of waterbar in situ. This is a promising line of approach to the sealing of water-retaining structures and although the cost of these compounds is prohibitive at present, the enhanced performance and reduction in labour costs merit serious consideration.

Natural and synthetic rubber. Rubber has been used to seal joints and cracks in concrete structures in a number of different ways. Rubber strips have been clamped to the surface of the concrete by means of steel plates and bolts so as to span the joints in water-retaining structures and recently[29] a reservoir which had been seriously fractured by soil subsidence was completely lined with a preformed rubber membrane. The most common use of rubber however is in the form of a valve type of waterbar developed by the Bureau of Reclamation, Denver, Colorado, in the 1930's. Originally vulcanized natural rubber was used for the manufacture of these waterbars but developments in the production and use of plastics have led to the use of various long chain polymers having rubber-like characteristics.

For the sake of simplicity it is desirable to consider rubber waterbars in three groups according to the properties and particular uses of the materials used in their manufacture.

Natural rubber. Although it is an established fact that rubber deteriorates when exposed to air and sunlight, evidence is available[30] of rubber gaskets which after recovery from water and gas mains after sixty years service gave mechanical rest results of the same order as would be expected from new materials. The conditions under which rubber waterbars are used in concrete structures should, if anything, be less severe than those in gas mains in particular and with the improved methods of compounding rubber in present-day use, an effective life of more than 60 years may be expected from rubber waterbars.

Typical profiles of natural rubber waterbars, most of which are the original designs are shown in Fig. 8. 4. These profiles are produced either by moulding or by a straightforward extrusion technique. In general, moulded waterbars will have a more uniform profile than the extruded types and there is greater freedom of choice of the composition of the rubber according to the desired characteristics of the finished products.

Natural rubber can be compounded with fillers, wax and vulcanizing materials to produce a material possessing to a high degree most of the properties required in a waterbar. The ability of rubber waterbars to accommodate differential settlement of adjacent sections of concrete, which is probably the most severe type of movement, was demonstrated in 1936 when deflections of up to six inches were accommodated at joints in the All-American Canal[31]

100

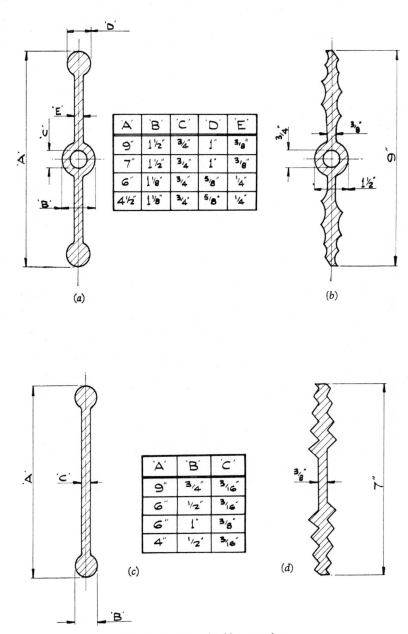

FIG. 8. 4. Natural rubber waterbars.

FIG. 8. 5. (a).
Deformation of a natural rubber waterbar during a water pressure test. Above, at start of test. Right above, at extension of 9.5 inches, below, after test.
Photographs by permission of Electricité de France

Fig. 8. 5. (b).

103

without any deleterious effects. These waterbars will also withstand considerable water pressure even when this is accompanied by extension. A striking example of combined water pressure and extension is shown in Fig. 8. 5[32]. This photograph was taken during a small scale trial in which water pressure was applied to a hollow cylinder of concrete having a rubber waterbar in a contraction joint at the mid point of the longitudinal axis. Extension of the waterbar was allowed to occur until at an elongation of $10\frac{1}{4}$ in. and a water pressure equivalent to a 200 ft. head the rubber fractured. No leakage of water occurred during this test !

The pronounced elastic behaviour of rubber makes it particularly suitable as a valve type waterbar and its deformability ensures that close contact is maintained between the anchorage points and the concrete to provide an effective seal.

The chemical resistance of natural rubber is adequate for most civil engineering applications, but it should not be exposed to oils or organic solvents.

Rubber waterbars are usually installed by the use of split wooden shuttering and a suitable clamping device. Apart from this small complication in formwork construction, however, the installation of these waterbars is relatively simple due to the ease with which they may be bent. Vertical curves of 6 in. radius and horizontal curves of 30 ft. radius *are* possible with the heavy nine inch wide section. Prefabricated networks of the waterbar can normally be obtained or joints at intersections may be vulcanized on the site using equipment which is supplied by the manufacturer.

Polyvinyl chloride. This is a thermoplastic material which is manufactured from simple raw materials and is therefore one of the cheapest synthetic plastics. By the addition of softening agents known as plasticizers its flexibility may be varied over a very wide range and being extremely " rubbery " it has become a natural development from the use of natural rubber as a waterbar material.

Polyvinyl chloride, or PVC as it is commonly known, is easily manipulated and may be moulded or extruded with great precision. In addition therefore to the relatively simple profiles of the natural rubber waterbars a variety of other profiles have been produced, usually with the intention of restricting the flow of water by providing a long and tortuous path between the waterbar and the concrete surrounding it. A number of these profiles are shown in Fig. 8. 6.

Whereas the flexibility of PVC can match that of rubber it is not quite as elastic and therefore has a slower rate of recovery after deformation. Its tensile strength is also lower than that of good quality

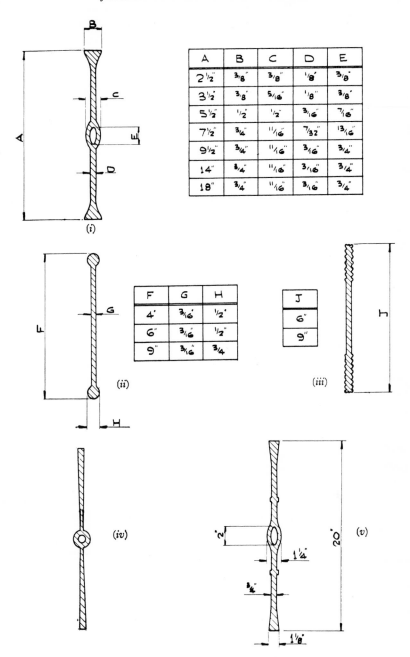

A	B	C	D	E
2½"	3/8"	3/8"	1/8"	3/8"
3½"	3/8"	5/16"	1/8"	3/8"
5½"	1/2"	1/2"	3/16"	7/16"
7½"	3/4"	11/16"	7/32"	13/16"
9½"	3/4"	11/16"	3/16"	3/4"
14"	3/4"	11/16"	3/16"	3/4"
18"	3/4"	11/16"	3/16"	3/4"

F	G	H
4"	3/16"	1/2"
6"	3/16"	1/2"
9"	3/16"	3/4

J
6"
9"

(i)

(ii)

(iii)

(iv)

(v)

FIG. 8.6. Valve type PVC waterbars.

105

rubber and it is more susceptible to variations in temperature. Considering the rate and magnitude of the movements occurring in concrete structures however the properties of PVC are almost certainly more than adequate to fulfill the functions of a waterbar.

PVC has the very significant advantage over other waterbars of the valve type in that it can be autogeneously welded by a simple heat treatment. It is possible therefore for all types of intersection to be made on the site.

The resistance of PVC to weathering and chemical attack is generally very good but on contact with oils migration of the plasticizers may occur causing the material to become hard and brittle on drying out. An interesting new development in PVC waterbars is a composition which is resistant to gamma radiation and therefore suitable for use in structures exposed to radio-active materials such as atomic power stations.

A very good fatigue life may be expected in a PVC waterbar particularly if the plasticizer content is high ; fatigue life is significantly reduced by the introduction of fillers.

Synthetic rubber. This term is often applied to all synthetic thermoplastic materials having elastic properties including materials such as PVC which possess elasticity but have a distinctly different chemical structure from natural rubber. A number of synthetic thermoplastic materials, however, have structures and properties very similar to those of natural rubber and can reasonably be classified as synthetic rubbers. These materials can be vulcanized and compounded with carbon black and other fillers which are normally used in natural rubber to obtain optimum performance. Their chief advantages over natural rubber for use as waterbars are resistance to oxidation and ageing and resistance to oils and organic solvents. In other respects, such as temperature susceptibility and elasticity, their properties are inferior to those of natural rubber but the same considerations apply here as in the case of PVC, namely, that these properties greatly exceed the requirements of a waterbar.

Being vulcanized, synthetic rubber waterbars are jointed in the same manner as natural rubber and in this respect they are less convenient than thermoplastics such as PVC.

WATERBAR DESIGN.

A wide variety of waterbar designs is available at the present time not only in metals and poured materials which can be fashioned on the site, but also in manufactured materials such as rubbers and PVC.

The important features requiring consideration when choosing a waterbar for a particular structure are that it should :—

(a) be adequate to accommodate the conditions to be encountered in service.

(b) utilize the properties of the material of which it is made to the greatest advantage.

(c) be as simple as possible in order to promote ease of installation.

Metallic waterbars. The various metals which have been considered for use as waterbars function either as valves or by virtue of the close contact or bond which is made when the concrete is placed against the waterbar. These two methods of providing a seal require entirely different designs in order to give the best possible performance. Whereas the first method depends on stress concentrations at the points of anchorage and therefore throughout the diaphragm the second can operate effectively only if the centre of the waterbar is sufficiently flexible to prevent the development of stresses which will cause the bond between the concrete and the waterbar to fracture.

In practice however a compromise must be made between these two ideal sets of conditions. This compromise can be applied according to the magnitude of the movement anticipated at the joint. If the movement to be accommodated is very small e.g., such as would be expected at partial contraction joints, a valve type of waterbar similar to (a) in Fig. 8. 3. which relies on the strength of the metal could be chosen. Under these conditions also a waterbar such as (d) in Fig. 8. 3. relying on bond strength and installed so as to permit flexibility of the central portion would be expected to perform satisfactorily. The introduction of holes into this type of waterbar as in (b), whilst preventing displacement of the diaphragm embedded in the concrete, also shortens the path for the passage of water through the joint and is probably of little advantage.

Where larger movements are anticipated, however, a compromise between these two designs must be made. The design (j) in Fig. 8. 3. is a typical example of a metallic waterbar intended to accommodate movement. This waterbar has a relatively small area of contact with the concrete at the anchorage points so that high stress concentrations develop at these points although there is considerable allowance for movement at the centre of the waterbar. Many waterbars of this type has been used and are described in detail elsewhere[33].

Probably the most effective use of metallic waterbars is in conjunction with poured waterbars such as bitumen.

Poured waterbars. The design of cross-section of poured waterbars is governed largely by the ease with which the cavity to hold the waterbar can be formed. The most common shapes which have been used

107

therefore are circles, triangles, squares and diamonds because all of these permit the shuttering to be struck easily. Wedge-shaped cross-sections have been used occasionally but not only is this shape unlikely to affect performance, it also requires complicated shuttering to form the cavity and difficulties are experienced in striking the shuttering when the concrete has set.

As indicated in an earlier section in this chapter two of the most serious shortcomings of poured waterbars are :—

(a) the difficulty in ensuring that the material penetrates into the cavity to fill it completely.

and (b) the tendency of the material to flow through the joint due to gravity or water pressure.

The first of these can be mitigated in the case of heated materials by installing a steam heating tube in the form of a U at the centre of the cavity. This tube may then be slowly withdrawn whilst the material is still molten. If adequate clearance is provided to accommodate movement of the concrete the tube may be left in position to render the material fluid whilst the structure is in service. This may serve as a remedial treatment when the material has lost adhesion to the concrete. In view of the damp conditions which will exist in the joint in such an event, however, it is unlikely that this form of treatment would be very effective.

The tendency for a poured waterbar to flow through the joint can be overcome by installing a metallic waterbar across the joint either on both sides of the poured waterbar or on the side remote from the anticipated water pressure. This combination of poured and metallic waterbars has been used extensively in all types of water-retaining structure and if the materials are properly installed, satisfactory performance should be obtained.

Natural and synthetic rubber waterbars. These waterbars with the possible exception of (b and d) in Fig. 8. 4. operate on the valve principle and thus the four main features to be considered in their design are as follows :—

(a) The profile should be simple in order to permit the concrete to be placed and compacted easily. It is essential to obtain the maximum area of contact between the concrete and the waterbar. Simple profiles may also be jointed easily and efficiently.

(b) The distance between the valves or points of anchorage should be sufficient to ensure that dense waterproof concrete can be obtained around them. In no case should this distance be less than the depth of cover provided for the reinforcement and for preference it should not be less than the maximum size of aggregate used in the concrete.

(c) In order to obtain the maximum possible concentration of stress at the anchorage points the distance between these points should bear the correct relationship to the thickness or modulus of elasticity of the material between them.

(d) The possible area of contact between the concrete and the waterbar at the points of anchorage should be large enough to provide adequate allowance for the small imperfections in the degree of contact caused by small bubbles of entrapped air. The possible area of contact should not, however, be so large as to significantly reduce the stress concentration between the waterbar and the concrete.

Most of these requirements appear to be satisfied by the original designs developed by the Bureau of Reclamation, (c) being suitable for joints to accommodate simple expansion and contraction effects and those with hollow central bulbs (a) being designed to accommodate differential movement between abutting sections of structure.

PVC Waterbars. The general requirements for the design of PVC waterbars are the same as those for rubber. PVC is however more easily manipulated than rubber so that in addition to the simple profiles ((ii) in Fig. 8. 6.) used in valve type waterbars, more complicated profiles (as shown in Figs. 8. 7. and 8. 8.) have been produced which are intended to resist the flow of water by providing a long tortuous path across the joint. Some designs combine these two methods of providing a seal.

In combining the two methods optimum performance will be achieved when the projections acting as valves are given adequate concrete cover and the profile of the waterbar between them is smooth so that the stress which develops due to shrinkage of the concrete is concentrated at the valves. Three designs employing these principles are (d) in Fig. 8. 7., (f) in Fig. 8. 8., and Fig. 8. 9. The last of these combines all the desirable features of valve type waterbars.

A method of sealing the horizontal lift joints in walls which the author considers worth investigating is the use of PVC or rubber tubing. It seems feasible that PVC tubes could be produced in a range of sizes, variations being made in both external and internal diameters such that for any given loading due to the overburden of concrete, a size of tube can be chosen which will deform elastically and thus be maintained in a stressed condition. A waterbar of this design should be economical both in material and manufacturing costs and would also be relatively easy to install and joint.

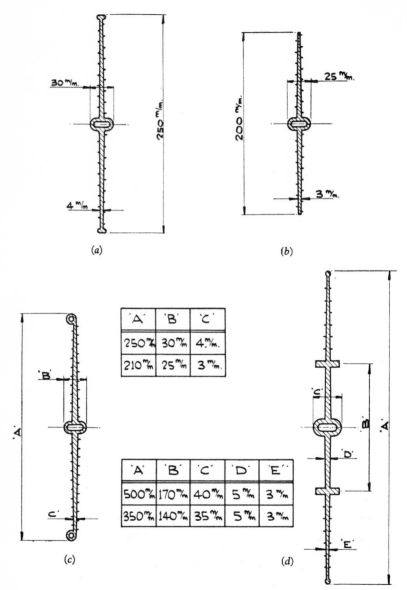

A	B	C
250 ᵐ/ₘ	30 ᵐ/ₘ	4.ᵐ/ₘ.
210 ᵐ/ₘ	25 ᵐ/ₘ	3 ᵐ/ₘ.

A	B	C	D	E
500 ᵐ/ₘ	170 ᵐ/ₘ	40 ᵐ/ₘ	5 ᵐ/ₘ	3 ᵐ/ₘ
350 ᵐ/ₘ	140 ᵐ/ₘ	35 ᵐ/ₘ	5 ᵐ/ₘ	3 ᵐ/ₘ

Fɪɢ. 8. 7.   Ribbed PVC waterbars.

110

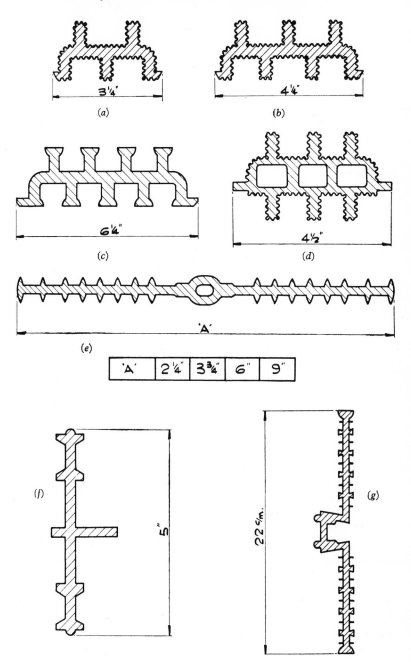

| 'A' | 2¼" | 3¾" | 6" | 9" |

Fɪɢ. 8. 8.   Other designs of PVC waterbar.

111

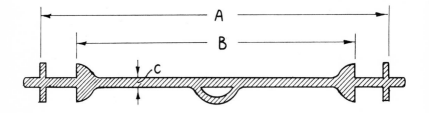

'A'	'B'	'C'
4"	2 ¾"	⅛"
6"	4 ½"	⅛"
8"	6 ¼"	³⁄₁₆"
9"	7"	³⁄₁₆"
10"	7"	³⁄₁₆"
10"	8"	¼"
11"	9"	¼"

FIG. 8. 9.   A double valve P.V.C. waterstop.

CHOOSING A WATERBAR.

No definite recommendations can be given on the choice of waterbar designs or materials, either for general, or for specific types of construction because experience has shown that various · materials and designs can give satisfactory performance under widely differing circumstances.

The characteristics considered by the author to be desirable in waterbars for a number of selected types of structure based on ease of installation and the important features of waterbars of the two main types are indicated in Table 8. 1. Examination of these characteristics together with the designs and properties of materials available will, it is hoped, provide a general guide to the most suitable combination for a specific purpose. Obviously installation requirements on the site will in some cases conflict with functional requirements of the waterbar and in these cases a compromise may be necessary.

112

TABLE 8. 1. A SUMMARY OF THE IMPORTANT CHARACTERISTICS REQUIRED IN WATERBARS FOR VARIOUS TYPES OF STRUCTURE.

Type of Structure or joint		Waterbars relying on adhesion or length of path.			Valve Type Waterbars		
		Flexibility	Width	Modulus of Elasticity	Flexibility	Width	Modulus of Elasticity
Water-retaining structures in average conditions	Expansion joints	good	large	high	good	fairly large	high
	Shrinkage or contraction joints	fair	fairly large	low	fair	small	low
	Construction or partial contraction joints	not important	large	not important	not important	small	high
Structures on foundations of dubious bearing properties		not recommended			very good	large	low
Dams and other large mass concrete structures		not important	large	high	fair	fairly large	high
Heavily reinforced structures		good	small	not important	good	small	high
Buildings		fair	fairly large	not important	not important	small	high

H

113

CHAPTER NINE

Joint Design

The design of joints must be based on the following conditions :—
(a) The type of structure.
(b) The method of construction.
(c) The function of the joint.
(d) The jointing materials available.

The most important single variable which is also the most convenient basis for reference is the type of structure and joint design will therefore be considered in relation to types of structure. It will be possible in the scope of this book to comment in detail on only a few typical joint designs which are in common use. The desirable and undesirable features of each type will, however, be covered as fully as possible.

WATER-RETAINING STRUCTURES.

The design and construction of the joints in a water-retaining structure are often of vital importance to the satisfactory performance of the structure and the economic use of the undertaking as a whole. Poor joint design or inadequate spacing of joints to relieve stresses in the concrete may involve expensive pumping costs and in addition progressive deterioration of the structure may occur which will eventually necessitate partial or complete reconstruction. The design of joints in water-retaining structures can be considered according to the type of construction as follows :—

Mass concrete. The effect of shrinkage of the concrete during setting is of particular importance in mass concrete structures and to minimize the effects of shrinkage the alternate bay system of construction is most common. Very large structures such as gravity dams are often constructed with small closure sections which are placed after initial shrinkage has taken place in the rest of the structure. This method of construction doubles the number of joints but reduces the amount of extension to be accommodated by the materials used to seal these joints. The properties required in the sealing materials, therefore, are alleviated to some extent.

When adjacent sections of a structure such as a reservoir or dam interlock by means of a tongue and groove, shrinkage or contraction must be minimized in order to prevent the joints opening and thus reducing the lateral stability. Usually such structures are in permanent compression.

(i) *Expansion joints.* These are not usually provided in mass concrete water-retaining structures because they detract from structural stability. In these structures expansion of the concrete is largely offset by initial shrinkage and any stress which may develop due to expansion will assist the structure to resist water pressure. The only structures likely to require expansion joints are sea walls, flood relief channels and other similar structures which may be fully exposed to atmospheric temperatures for long periods during the summer.

There is little advantage in providing a tongue and groove in expansion joints as shown in Fig. 9. 1. (e). These joints are filled with a readily compressible material and thus considerable deflection of the concrete can occur before any load is transferred from one section to the other. The tongue and groove must have parallel sides as shown in (f) in order to transfer load. An alternative method is to use some form of dowelling such as those illustrated in Fig. 9. 5. for joints in pavings.

(ii) *The position of the waterbar.* A waterbar is usually incorporated in the joints in mass concrete structures in order to seal cracks which develop due to shrinkage. The waterbar should be placed at a sufficient depth below the water face to permit the concrete between it and that face to be placed and compacted easily. It is suggested that this depth be not less than three times the maximum size of aggregate used in the concrete or half the width of the waterbar, whichever is the greater.

When a tongue and groove joint is used the waterbar should not be placed within the tongue and groove as shown in Fig. 9. 1. (g) because this creates an undesirable plane of weakness and can lead to structural failure even if a thin metal diaphragm is used[34]. A more suitable position for the waterstop is shown in (a), (d) and (f).

Joint designs incorporating copper waterbars should include space for a bituminous or compressible material where the waterbar intersects the joint. A second cavity may also be provided for subsequent filling with bitumen to form a second line of defence. Two typical arrangements are (b) and (c) in Fig. 9. 1. In (b) where no closure section is employed to reduce shrinkage effects two copper strips are installed to prevent the bitumen flowing out of the joint when

115

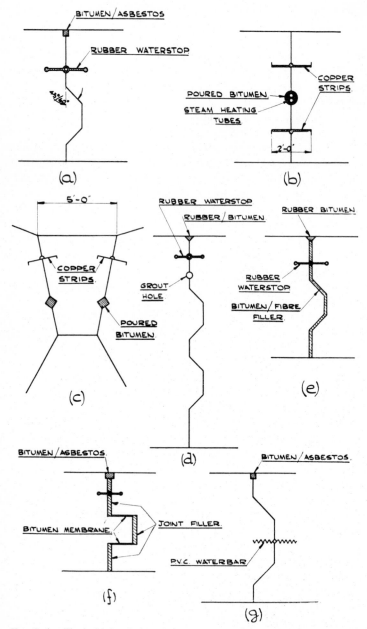

FIG. 9. 1. Typical joint designs for mass concrete water-retaining structures.

it opens. The use of the closure sections in (c) reduces the tendency for the joints to open and under these circumstances it may not be necessary to take preventative measures against possible displacement of the bitumen.

Reinforced concrete. The need or desirability of incorporating shrinkage and expansion joints in reinforced water-retaining structures depends on :—

    (a) the general principle of the design.

    (b) the degree of exposure.

    (c) soil conditions.

(i) *Monolithic structures.* Some reinforced concrete structures are designed on the " box " principle in which a high steel/concrete ratio is used and reinforcement is continuous through the floor, walls and roof. In this type of construction the concrete must have a low water/cement ratio in order to minimize shrinkage. The thickness of the concrete should also be as small as possible so as to reduce differential volume changes. It is difficult to reconcile these two considerations with the task of placing and compacting the concrete, particularly in view of the amount of reinforcement required to accommodate tensile stresses. Very close control is necessary therefore in this type of construction to ensure the production of uniform stress conditions throughout the structure. Construction joints should be placed at sections of minimum shear and particular attention must be paid to the preparation and treatment of these joints.

(ii) *Articulated structures.* The difficulties that arise in the design and construction of monolithic structures to ensure satisfactory performance under service conditions which are rarely completely predictable can largely be avoided by providing some degree of articulation. A few typical designs for joints in reinforced structures are shown in Figs. 9. 2. to 9. 4. These joints may be considered in detail as follows :—

Expansion joints. These joints will be required in those parts of the structure exposed to atmospheric conditions and in structures built on soils having poor or unpredictable bearing properties. When atmospheric conditions only will be experienced the types of joints shown in Fig. 9. 2. (b), (d) and Fig. 9. 4. (i) should be suitable. The joint shown at (b) is applicable to both floors and the vertical joints in walls if the appropriate type of sealing compound is used. The essential features of these designs are as follows :—

    (a) A flexible rubber diaphragm having no degree of freedom may be used.

    (b) Where possible a floor slab should rest on the wall footings and the base slab, not the wall on the floor.

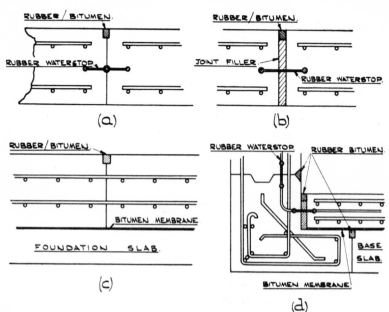

FIG. 9. 2.   Typical joint designs for the floors of reinforced concrete water-retaining structures.

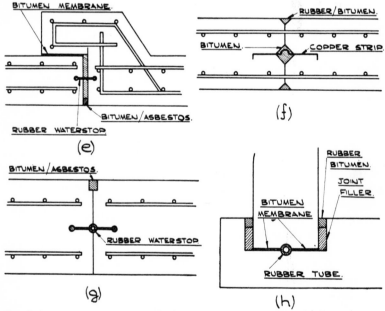

FIG. 9. 3.   Typical joint designs for the walls of reinforced concrete water-retaining structures.

(c) A PVC waterbar should have a degree of freedom provided by a hollow centre bulb to increase its ability to accommodate movement without permanent deformation occurring.

(d) No filler or sealing compound is required in the underside of roof joints. This permits condensation to escape easily and dispenses with the difficult task of applying a sealing compound to these joints.

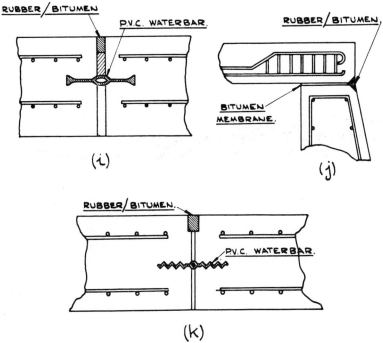

FIG. 9. 4. Typical joint designs for the roofs of reinforced concrete water-retaining structures.

If soil movements are anticipated, larger sections of rubber diaphragm as shown in Fig. 9. 2. (a) should be used in expansion joints in place of the parallel sided diaphragm unless some load transfer device is provided across the joint as in Fig. 9. 3. (e).

Shrinkage or contraction joints. Typical designs for joints to permit free shrinkage of the concrete in the floors and walls of reinforced structures are shown in Fig. 9. 2. (a) and Fig. 9. 3. (g). These joints provide a complete break in the structure and may therefore accommodate warping due to differential volume changes. It is essential that the waterbar spanning these joints be allowed a degree of freedom and this is usually provided by a hollow central bulb.

Occasionally[35] a deformable material such as bitumen is provided to permit lateral movement but some interaction between the waterbar and the bitumen might occur and lead to deterioration of the water-bar.

In view of the temperature variations which occur in roofs, shrinkage can usually be accommodated by the joints provided to permit expansion. When the roofs are insulated from atmospheric conditions soon after construction, however, the expansion joints may not be placed at spacings close enough to accommodate shrinkage. In such cases contraction or partial contraction joints of the types shown in Fig. 9. 2. (a) and Fig. 9. 4. (k) respectively should be installed.

If large soil movements are anticipated expansion joints should be installed at spacings of about 30 ft. and then they also accommodate shrinkage of the concrete.

Partial contraction joints. These joints are installed when heavy reinforcement is used to give a structure greater stability than may be achieved with a fully articulated design. Typical designs for partial contraction joints are shown in Fig. 9. 2. (c), Fig. 9. 3. (f) and Fig. 9. 4. (k). Movement at these joints is restricted by the reinforcement and waterbars of the valve type should not be so readily deformable as to prevent the development of high stress concentrations at the points of anchorage. Waterbars which rely on adhesion or length of path across the joint can be expected to give satisfactory performance, particularly if they are designed or installed in such a way that the small movement which may occur is accommodated at the centre of the waterbar.

" *Lift* " *joints.* The most commonly used type of lift joint, which is installed between successive layers of concrete, is shown in Fig. 9. 2. (d). The waterbar used in this type of joint should satisfy the same requirements as those applying to the waterbar in partial contraction joints.

A design which in the opinion of the author merits consideration, particularly in structures such as circular prestressed tanks which have no vertical joints is shown in Fig. 9. 3. (h). This design is based on the use of a waterbar of hollow rubber or PVC tubing, the diameter and wall thickness of which are chosen to suit the loading conditions imposed on it by the upper layer of concrete. This type of waterbar has been considered under the heading " PVC water-bars " in Chapter Eight.

Wall/roof joints. The most reliable type of joint between the roof and the walls of an articulated concrete structure is the sliding joint (j) in Fig. 9. 4. This joint should be constructed with great care to ensure that the roof slab is free to slide over the walls. The top of the walls should be floated level and smooth during construction in order to prevent a mechanical key and bond between the two sections of concrete should be prevented by the insertion of an incompatible membrane. Hot or cold-applied bitumen is often used as an isolating medium and providing that the tops of the walls have been finished correctly this treatment is adequate. A more effective membrane which makes some allowance for irregularities in the upper surfaces of the walls is lead, copper or aluminium cored damp-proof course material. The extra expense of using such a material is often justified by the somewhat greater tolerance permissible in the finish of the walls.

SMALL WATER-RETAINING STRUCTURES.

Structures which impound small heads of water, such as swimming baths, paddling pools and various types of treatment tank and effluent channel in sewage works, will not usually require expansion joints except, possibly, to isolate the walls from the floor. Suitable joint designs for these structures are (a), (d), in Fig. 9. 2. and (e) and (g) in Fig. 9. 3. using the lighter sections of waterbar. The waterbars used in these joints should, as a rule, be capable of accommodating lateral movement because it is difficult to form a tongue and groove in thin sections of concrete.

PAVINGS.

It has been suggested in Chapter Five that, on the basis of full-scale experience, pavings may be constructed without joints of any kind. A survey[36] of current practice, however, has revealed that two general principles are being used in the design and construction of concrete roads. These principles can, with minor limitations, be extended to all other types of paving subjected to normal service conditions.

Partially articulated pavings. This heading will be used to cover pavings in which allowance is made for initial shrinkage but no joints are provided to accommodate expansion of adjacent slabs except at bridges or when two pavings intersect. This design is particularly suited to modern methods of construction using continuous laying and finishing equipment because it avoids the difficulty of installing a joint filler. It has been adopted for both plain and reinforced concrete but is likely to be more successful with the latter. Typical designs for joints to accommodate shrinkage are shown in Fig. 9. 5. (a), (b) and (c).

(i) *" Dummy " joints.* The " dummy " joint (a) comprises a

121

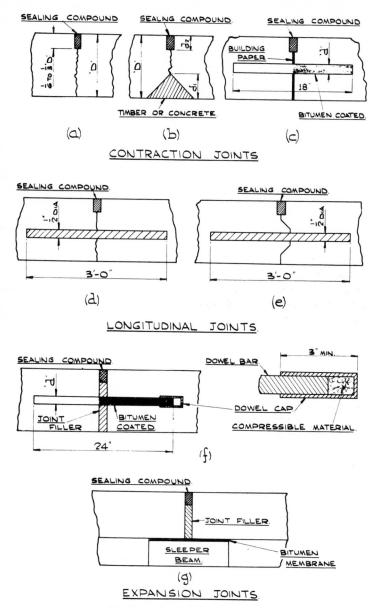

Fig. 9. 5.   Joint designs for pavings.

groove $\frac{1}{3}$ to $\frac{1}{5}$ of the depth of the slab which is either formed during construction or is cut in the concrete at some stage during the hardening process. This groove forms a plane of weakness so that as stress develops due to shrinkage, the concrete cracks at this plane and random irregular cracks in the paving are avoided.

Reinforcement should be laid continuously across dummy joints to provide some restraint on the tendency for the joints to open, once cracks have developed due to shrinkage. The reinforcement, placed in this manner, helps to ensure that some degree of load transfer is provided by the irregularity of the crack and is capable itself of transferring load across the joint to a limited.extent.

In another design of " dummy " joint (b) the depth of the groove at the surface of the paving is reduced by providing a plane of weakness immediately below it at the base of the slab. This design is more economical in that it reduces the volume of sealing compound required and it also reduces the sawing costs when the surface groove is formed by this means.

(ii) *Construction/contraction joints.* It is necessary to form a joint at the end of a working period. It is desirable that this joint be made a deliberate break in the paving to avoid the possibility of random cracking in adjacent slabs in the event of a bond developing between them. The most convenient method of constructing this break is to turn up the paper underlay against the stunt end and to retain it at the top with a timber fillet which is used for forming the sealing groove. As reinforcement cannot conveniently be carried across this type of joint steel dowel bars should be set into the concrete. The dimensions of these dowel bars should be the same as those given in Table 9. 1. for expansion joint dowels but the cap containing compressible material may be omitted.

(iii) *Longitudinal joints.* Concrete pavings wider than fifteen feet should be provided with longitudinal joints at about 12 ft. spacings to accommodate warping of the concrete and possible soil movements. Two types of longitudinal joint are shown in Fig. 9. 5. (d) and (e); the former is used when the width of the paving machinery exceeds fifteen feet and the latter is for use between successive lanes. Tie bars or continuous reinforcement should be provided at longitudinal joints for the following purposes :—

(i) To prevent the joints from opening.
(ii) To transfer load thus ensuring equal deflections of adjacent slabs.
(iii) To preserve a level surface should soil movements occur.

When reinforcement only is used it may be necessary to increase the size of the transverse bars in the fabric.

TABLE 9. 1.[37] Minimum diameter of dowel-bar required to withstand failure under various conditions (for a joint having a load-transfer efficiency of 80 per cent.)

Type of paving		Spacing of dowel-bars inches.	Modulus of subgrade reaction : lb./sq.in./in											
			50				200				500			
			Thickness of concrete slab : inches											
			6	8	10	12	6	8	10	12	6	8	10	12
Air fields	Wheel load lb.													
	25,000	12	–	3/4	3/4	3/4	–	3/4	1/4	1/4	1	3/4	3/4	3/4
	50,000	12	–	–	3/4	3/4	–	1	1	1	–	1	1	1
	75,000	12	–	–	1	1	–	–	1	1	–	–	1 1/4	1
Roads	Joint width in.													
	0	12	1/2	1/2	1/2	1/2	3/4	3/4	3/4	3/4	3/4	1	1	1
	1/2	12	1/2	3/4	3/4	3/4	3/4	1	1	1	1	1 1/4	1 1/4	1 1/4
	3/4	12	3/4	3/4	3/4	3/4	1	1	1	1	1 1/4	1 1/4	1 1/4	1 1/2
	1	12	3/4	3/4	3/4	3/4	1	1	1 1/4	1 1/4	1 1/4	1 1/2	1 1/2	1 3/4

– indicates thicknesses of slab likely to be insufficient for the load to be carried.

Note : The moduli of subgrade reaction cover the following soil types :—

50 — Clays of high plasticity and organic silts.
200 — Low plasticity clays and silts and poorly graded and fine sands.
500 — Well graded and clayey sands and gravels.

Various types of tie bar have been used including deformed reinforcing rod, and hooked bars which are linked at the joint. Providing the length is adequate, however, plain bars at a spacing of two feet along the joint will generally be found satisfactory in the two joint designs shown. If deformed rod or hooked ends are used the length of the tie-bars may be reduced and if a straight butt joint is constructed, the diameter of the bars should be increased to $\frac{3}{4}''$ for medium and to one inch for heavy loading conditions.

Fully articulated pavings

The most common design for concrete pavings in use at the present time provides joints to accommodate both contraction and expansion of the concrete.

Nearly all the concrete pavings constructed prior to 1939 were provided with expansion joints at spacings of up to 60 feet unless they were used as a foundation for a bituminous surfacing when construction joints only were sometimes employed. The development of continuous laying machinery, however, which has now superceded hand laying equipment on all but the very small jobs has led to drastic changes in the design of pavings.

When concrete is laid by modern continuous placing and finishing equipment the installation of expansion joints not only complicates construction, but also has a deleterious effect on the regularity of the finished surface. The present trend, therefore, is to restrict the number of expansion joints to the minimum considered necessary to prevent the development of dangerous stresses in the concrete due to changes in moisture content and temperature. Thus although some pavings are still constructed with expansion joints at spacings up to 60 feet the majority of fully articulated pavings are constructed with both expansion and contraction joints at intervals of the order given in Table 2. 1.

The use of expansion joints at large spacings with intermediate contraction joints, however, leads to difficulties in the design and treatment of joints for the following reasons :—

(a) Contraction joints tend to open and the transfer of load between adjacent slabs is reduced or prevented.

(b) Very large movements may occur at expansion joints.

(c) The expansion joints become progressively narrow as the contraction joints open.

The possible advantages of incorporating expansion joints in a paving are that the weight of reinforcement may be reduced and the possibility of buckling due to differential expansion or soil movements is virtually eliminated.

The most common design of expansion joint in current use is shown at (f) in Fig. 9. 5. This joint incorporates steel dowel bars to transfer load across adjacent slabs thus ensuring uniform deflection. Half of the dowel bar is coated with a film of bitumen, grease or some other thin membrane to prevent the concrete bonding to it. A cap is also provided to cover this end of the bar, a wad of compressible material such as cotton waste being placed in the bottom of the cap to accommodate expansion of the adjacent slabs of concrete. Suggested minimum sizes and spacings for dowel bars for roads and airfields are given in Table 9. 1.[37]

It has been suggested that in order to permit expansion joints to open and close freely the maximum permissable tolerance in the alignment of dowel bars should be ± 0.10 inch in 12 inches for dowel bars up to and including $\frac{3}{4}$ inch and $\pm\ 0.05$ in 12 inches for larger diameters.

The performance of dowelled joints. In assessing the value of dowel bars in reducing the relative deflection between adjacent slabs the effects of temperature and moisture content gradients in the slab must be considered. When the temperature and moisture content of the concrete is lower at the top of the slab than at the bottom the edges of the slab will tend to warp upwards. This will reduce the contact between the base of the slab and subgrade and it has been shown[37] that large relative vertical deflection will occur when slabs

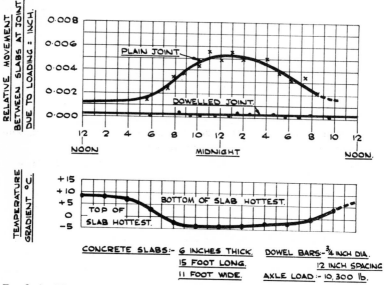

FIG. 9. 6.   The effect of dowel bars on the vertical deflection at the ends of paving-slabs.

in this condition are subjected to transient loading. These deflections will impair the riding quality of the paving and are likely in addition to cause premature failure of the sealing compound and spalling of the edges of the slabs due to impact.

The introduction of dowel bars does not reduce the warping of adjacent slabs significantly but as shown in Fig. 9. 6.[37] it does reduce by a considerable amount the vertical deflections occurring when the slabs are in a warped condition.

Sleeper beam joints. The use of a sleeper beam beneath expansion joints as shown at (g) in Fig. 9. 5., whilst having the advantage of providing a true and firm support for the jointing material during construction is not effective as a load transfer device for the following reasons :—

(a) It provides no load transfer when the edges of the slabs are warped upwards i.e., when the maximum deflections occur.

(b) Deflections of the edges of the slabs result in impact forces between the slabs and the beam which then tends to become embedded further into the subgrade. This will increase the deflections occurring in the warped and unwarped conditions.

The most effective use of concrete beneath an expansion joint is as a relatively thin layer of " lean " mix to provide a firm and level support for the jointing material and the dowel bars.

BUILDINGS.

It has already been suggested that it is necessary to provide only expansion joints in buildings except possibly in the floors where the use of light reinforcement results in the need to accommodate shrinkage. Sliding joints may also be necessary however, particularly at the junction of two members such as the roof and the walls. The usual and generally most effective method of accommodating expansion in a building is to provide a complete break in the structure from the foundations to the roof using double or split columns and beams.

Typical designs for joints in buildings are shown in Fig. 9. 7. The most important considerations in the designs are as follows :—

Walls above ground level. These may be designed to accommodate a surface sealing compound as in (a) or a jointing strip as in (b). There are advantages and limitations in both of these designs because whereas the light-coloured sealing compounds favoured for use in buildings will provide an efficient seal, the width of joint should be limited to $\frac{1}{2}''$. A jointing strip, on the other hand, can be used in wider joints but it may permit dampness to enter the building by capillary attraction and furthermore repeated flexing may cause it to harden and eventually fracture. However, providing that the joint spacing and width are chosen to suit the optimum characteristics of the two sealing media, long satisfactory performance can be expected with both designs. When a jointing strip is used good clearance should be provided between the corrugation and the walls of the joint to encourage water shedding.

It is not usually necessary to seal the joints in internal walls and the best treatment for these is a cover strip or an insert of compressible PVC. PVC inserts are particularly suitable for joints in internal walls because as they are installed prior to the building being put into service,

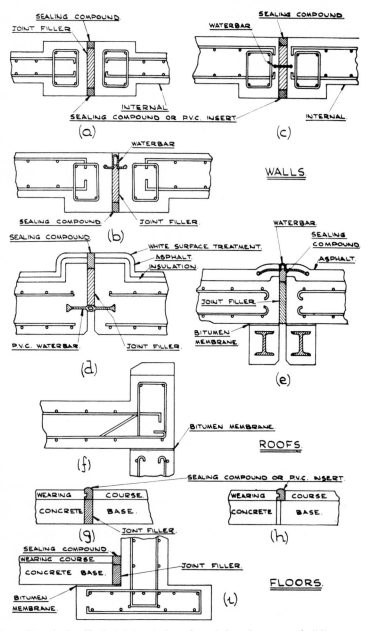

FIG. 9. 7. Typical joint designs for reinforced concrete buildings.

they are likely to be maintained in a state of compression by expansion of the concrete due to the subsequent higher level of air temperature under service conditions. These inserts may be used in addition to a sealing compound to enhance the appearance of external joints.

Walls below ground level. An efficient water barrier is essential in these walls particularly if the natural water table in the soil is near the surface. Frequently it will be impossible to provide an external surface seal to these walls and such a seal cannot in any case be provided in the floors. A waterbar should therefore be installed as shown in (c). In view of the restricted space available to accommodate a waterbar, the light PVC sections are most appropriate. These waterbars have the additional advantages of being easily jointed and bent to small radii.

Roofs. The provision of well designed joints at adequate spacings in the roofs of buildings is particularly important. The dimensional changes occurring in flat concrete roofs are often five times those experienced at foundation level due to the greater degree of exposure. It has been stated[38] that a flat concrete roof 50 ft. long can undergo a movement of about $\frac{3}{4}''$ due to variations in temperature. The difference between movements of this order and the relatively small movements occurring in the walls can cause serious distortion and cracking.

Roof treatments. The temperature variations in a roof can be reduced considerably by the provision of a layer of insulating material above the roof slabs. A sound absorbent material such as wood wool used as permanent shuttering on the underside of the roof will also assist in the thermal insulation of the structure. Further protection may be provided by the application of a reflective treatment to the surface of the waterproofing membrane which is usually laid over the insulating material. Suitable treatments are a coat of limewash,[38] a surface dressing using white chippings, cellular or no-fines concrete or the application of one of the proprietary synthetic resin materials designed for this purpose.

(i) *Expansion joints.* Expansion joints in roofs are often spaced at 100 ft. and although the application of protective treatments to the roof will go a long way towards reducing the amount of movement occurring due to variations in atmospheric temperature and solar radiation, considerable movement may still occur at these joints. Moreover the need for an effective seal against the penetration of water makes the selection of a good joint design and efficient jointing materials essential. It is desirable therefore to install a waterbar as an additional seal to a surface sealing compound.

129

Two typical designs for expansion joints in roofs are (d) and (e) in Fig. 9. 7. In (d) the waterbar is cast into the concrete thus permitting a layer of insulating material to be placed underneath the built-up coaltar or mastic asphalt roof surfacing. In (e) the waterbar is placed between successive courses of built-up asphalt membrane and the sealing compound is applied to a slot formed in the first course. The most important features which are common to these two designs are as follows:—

    (a) An upstand is formed on each side of the joint to encourage water shedding.

    (b) The membrane is not laid continuously over the expansion joint lest it prove incapable of accommodating the movements occurring at the joint without cracking.

    (c) The slot formed in the membrane to accommodate a sealing compound is continued to a depth of $\frac{1}{2}$ - 1 inch below the surface of the concrete.

    (d) When the joint filler is not supported by a waterbar, some form of mechanical fixing is provided.

    (e) The lower part of the joint may be left open, chamfers being formed at the edges to improve the appearance. It is possible, of course, to fill roof joints to the bottom but this leads to problems in the choice of decorative finish due to the movement to be accommodated. If an open joint is not acceptable a cover plate which is free to slide over the concrete at one side is probably the best solution.

(ii) *Roof/wall joints.* It is essential that roof slabs be free to move without causing distortion or displacement of the adjacent walls. The use of damp proof course materials has already been discussed.

(iii) *Floor joints.* The floor of a building may either be isolated from the walls using a joint of the type shown in (i) Fig. 9. 7. or the expansion joints in the walls may be carried through the floors using a design such as (g). In both methods intermediate construction joints of the type shown in (h) will usually be necessary. These construction joints will invariably open due to shrinkage effects and unless very heavy reinforcement is used to prevent this phenomenom, it is desirable to provide a cavity at the surface for a sealing compound or a PVC insert.

When a floor is isolated from the walls by means of an expansion joint a conventional sealing compound may be used with safety because it will not be subjected to traffic. Such a compound may in fact be necessary to accommodate the large movements which this joint may experience.

Expansion and contraction joints running across the floor are often subjected to severe and concentrated foot or vehicular traffic. It is essential that the slot be filled at the surface with a material which is capable of resisting this traffic and also of providing some support for the edges of the concrete. PVC inserts have been developed,[39] which, in addition to forming an arris, have a convex upper edge thus causing wheels to bridge across the vulnerable edge of the concrete. If it is necessary for the joints to be sealed against the ingress of water, these inserts may be bedded onto a sealing compound applied to the base of the formed slot. The elastomeric sealing compounds such as polysulphide and polysulphide/pitch are also particularly suitable.

THE CONSTRUCTION OF JOINTS

The methods adopted to construct the various types of joint in concrete must be chosen to suit the general construction programme. There are, however, a number of recognised methods of construction and these will be considered in relation to the general joint designs covered in this chapter. Construction joints have already been described in some detail in Chapter Three.

Water-retaining Structures. The most important requirements in the construction of joints in water-retaining structures are that :—
(a) a waterbar, when used, must be located to span the joint accurately. This is particularly important when a crimped copper waterbar or a flexible waterbar having a hollow central bulb is placed in a butt joint.
(b) the concrete surrounding the waterbar must be dense and waterproof.
(c) adequate clearance should be provided to permit concrete to flow easily between the waterbar and reinforcement. This clearance should be at least one inch except when the maximum size of aggregate is $\frac{1}{2}$ inch or less.
(d) when a cavity is formed for subsequent filling with a sealing compound the walls of this cavity should be regular, dense and waterproof.

(i) *Waterbars.* Waterbars can be installed most easily and efficiently by the use of timber end forms or a timber lining in steel forms. Malleable metal waterbars can be installed by bending the waterbar and fixing it to the end form as shown in (a) Fig. 9. 8. Another method which can be adopted with either metallic or flexible waterbars is the use of split shuttering as shown in (b) Fig. 9. 8. and Fig. 9. 9.

Attempts have been made to produce flexible waterbars which do not require the use of a split shuttering. One of these is shown in (c)

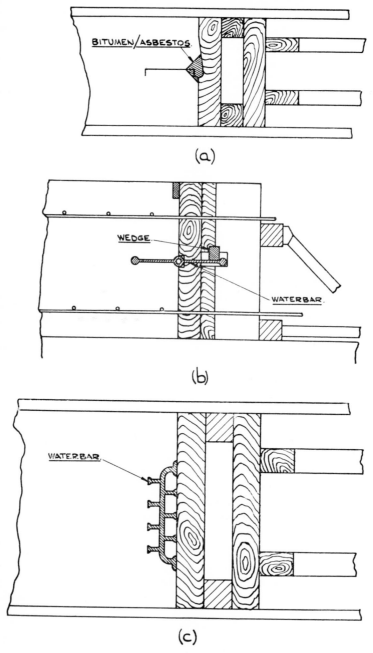

FIG. 9. 8.   Methods of installing waterbars.

Fig. 9. 8. but whereas this design of waterbar may provide a longer path through the joint it does not penetrate very deeply into the concrete and difficulties are likely to be experienced in obtaining dense water-proof concrete around it. The principle of fixing this waterbar to the form could, with advantage, be applied to both flexible and metallic waterbars by incorporating along the centre of the waterbar a continuous lug on each side thus enabling it to be nailed to the form. This would not only simplify formwork and but also provide an effective seal to prevent loss of fines through the split form. The loss of fines in this area must be prevented to avoid honeycombing around the waterbar.

Other features of waterbar design which assist in placing are the incorporation of a tying tag at the extremities of the waterstop and a bright pigmentation of the PVC compound so that the waterstop contrasts with the concrete thus assisting in accurate placing and

Fig. 9. 9. A rubber waterbar in split shuttering.

FIG. 9. 10.   The use of cross braces for locating a waterbar in a " lift " joint.

inspection. The design in Fig. 9. 9. features these particular aids to installation.

The installation of waterbars in horizontal lift joints may be carried out either by clamping the edge of the waterbar to conveniently placed reinforcement or by the use of cross-braces at intervals along the joint as shown in Fig. 9. 10.

Particular attention must be paid to obtaining fully compacted concrete in the region of a waterbar. When tamping by hand the concrete should be worked thoroughly underneath the waterbar in horizontal joints as shown in Fig. 9. 11. The appearance of the surface of the concrete at the junction of vertical and horizontal joints in walls is sometimes misleading during the compacting process. As compaction proceeds the excess water is brought to the surface and naturally collects at the vertical end form giving the impression that the concrete is fully compacted in this region. This may not be the case, however, and in addition to approaching as close as possible to the end form compacting effort should be continued in this area to ensure that the underlying concrete really is fully compacted. This will also minimize the possibility of an excess of weak high fines content mortar being formed in this very important area. This precaution must also be taken when placing the second section of concrete against a sliding joint so as to ensure that the sliding face is smooth.

(ii) *Reinforcement.* Adequate clearance must always be provided between reinforcement and the waterbar. The waterbar should never be bent or cut in order to prevent it fouling the reinforcement. If the joint is correctly located a reduction in the amount of reinforcement in the vicinity should be possible.

(iii) *The sealing cavity.* Appropriate formwork should be provided to enable a sealing cavity of the correct dimensions to be formed in dense waterproof concrete. It is not a good practice to use the joint filler to form the sealing cavity in expansion joints. Sealing cavities formed by the joint filler are usually irregular due to the material being compressed to varying degrees when the concrete is placed. Some joint fillers will also tend to absorb water from the concrete. This results in expansion of the filler and may also reduce the density of the concrete adjacent to it.

The sealing cavity should have a simple shape ; rectangular, triangular or tapered cavities can be formed easily and with reasonable accuracy. The shape of the cavity should be chosen to suit the type of sealing compound to be used to fill it as well as the conditions to which it will be exposed in service. It is important that the formwork is accurately constructed so as to prevent leakage of fines which would result in the walls of the sealing cavity being honeycombed.

135

FIG. 9. 11. Tamping concrete under a PVC waterbar.

Pavings. When simple methods are employed to place and compact the concrete, pavings are usually constructed by the alternate bay method. In this method of construction the shrinkage movement between adjacent slabs is approximately halved and as greater workability is required to permit the concrete to be fully compacted, a higher water/cement ratio is necessary. Considering these two factors it would appear that greater allowance should be made for movement due to temperature variations. It is customary, therefore, to adopt a close spacing for expansion joints in alternate bay construction, one or two intermediate contraction joints of type (c) in Fig. 9. 5. being provided to divide the paving into reasonable working lengths. All joints in this method of construction are formed against a stunt end as shown in Fig. 9. 12 (a). A timber fillet is provided to form the slot to accommodate sealing compound in contraction joints and in expansion joints the concrete is cast against a joint filler to which this fillet may be attached by metal clips. The edges of these joints should be rounded to a radius of $\frac{1}{4}$ - $\frac{3}{8}$ inch using an arrising trowel. Arrising should be carried out with care in order to avoid unnecessary irregularities adjacent to the joints.

(i) *Continuous construction.* When heavy laying and finishing equipment is used, pavings can be constructed in very large slabs, the only limitation being the capacity of the batching and mixing plant. Expansion, contraction and longitudinal joints may be formed in large slabs by two different methods. Both can produce greater regularity in surface finish than that obtainable with alternate bay construction. These methods are :

(a) The grooves to accommodate the jointing materials may be formed by vibrating a blade into the concrete immediately after it has been laid and compacted. The blade is allowed to penetrate to the depth required to provide a plane of weakness or to accommodate a joint filler. Vibration is then stopped, the blade is withdrawn and immediately a strip of joint filler is inserted into the groove to fill it completely. In order to distribute the humps which form due to displacement of concrete by the vibrating blade a straight edge 10 ft. long, attached to the end of a long flexible rod, is dragged transversely across the surface of the slab. In order to avoid displacement of the material in the joint and to ensure that mortar is not floated over this material there should be a rebate at the centre of the straight edge allowing it to bridge the joint. As a further precaution the filler should be slightly deeper than the groove so that it stands proud of the surface of the concrete. The complete process is illustrated in Fig. 9. 12. (b) (c) and (d).

When the concrete has hardened and cured, the joint filler may

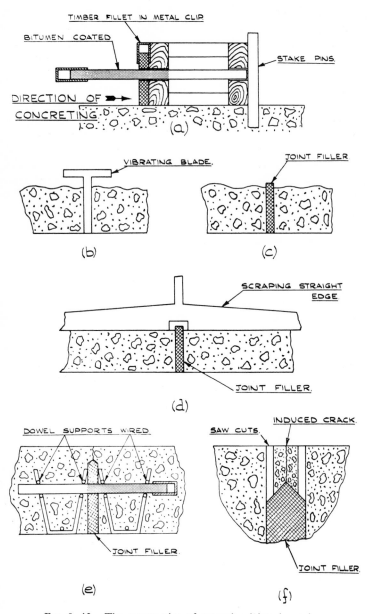

FIG. 9. 12. The construction of expansion joints in pavings.

be cut from the top of the groove to accommodate a sealing compound.

If dowel bars are used for load transfer purposes this method can be employed in the construction of " dummy " and longitudinal joints only.

(b)   An assembly carrying dowel bars, or the joint filler and dowel bars, may be set on the formation in advance of the concrete laying equipment.

In the construction of an expansion joint the joint filler in the assembly should have a depth slightly less than the total depth of the slab so as to permit concrete surcharge to pass over it with a minimum of backward surge.   A smaller strip of joint filler or a triangular wedge of timber or concrete may also be placed in a dummy joint assembly.   This reduces the depth of slot required at the surface to provide the necessary plane of weakness.   The concrete is then placed and finished continuously embodying the joint assemblies as shown in Fig. 9. 12. (e), and slots to accommodate sealing compound are cut in the concrete during the hardening process using diamond or carborundum-faced circular saws.

In this method of construction a very regular surface finish can be obtained but there are a number of points which merit close attention.   The most important considerations are as follows :—

(1)   A rigid assembly must be employed to locate the dowel bars and the joint filler accurately in expansion joints.   Several supporting devices have been used most of which are made from sheet metal, steel rod or reinforcing mat, the dowel bars being clamped or wired to the support at various points along their lengths.   These supporting devices are best used on a level bed of mortar previously laid on the formation.   A simpler method which has been used successfully is to place and vibrate concrete by hand methods on each side of the joint filler about one hour in advance of the main concreting work.   When this method is employed care must be taken to ensure that the concrete support is properly compacted.   It should not, however, be allowed to harden completely because this may have a deleterious effect on the bond between the support and the main slab.

The use of the main slab reinforcement as a support for the dowel bars is not entirely satisfactory because even with the aid of " chairs " it is difficult to ensure that the bars are accurately aligned.

(2)   The joint filler must be sufficiently robust to withstand the deposition and compaction of the concrete without tending

to fracture, buckle or fray.

(3) The position of the joint filler must be accurately marked. This is most conveniently achieved either by fixing a triangular fillet of timber to the top of the filler or by cutting the edge of the board itself to this shape. Treating the joint filler in this way induces a straight shrinkage crack in the concrete along the centre of the joint. The sealing slot may then be cut as shown in Fig. 9. 12. (f).

(4) " Dummy " joint grooves should be cut as soon as possible in order to avoid the development of an irregular crack. Attempts to cut these grooves too soon after the concrete is laid however will displace the large aggregate and cause spalling of the edges either during the cutting operation or subsequently. The appropriate time to cut these grooves will depend on weather conditions and will vary between twelve hours in warm weather and twenty-four hours or more in cold and damp weather. It has been stated [40] that concrete may be cut satisfactorily after it has reached a crushing strength of 1,000 lb./in.^{2} Cutting the grooves early results in economies in blade wear.

(5) The type of blade used to cut the grooves should be chosen according to the type of aggregate and the age at which cutting is carried out. When cutting is carried out at an early age or when soft aggregate such as limestone is used in the concrete, carborundum saws give the most economical performance. Harder aggregates such as granite and flint gravel may be cut economically with carborundum blades at early ages but when the concrete has hardened only diamond blades are satisfactory.

(6) The machines available for cutting grooves in concrete are of two types ; self-propelled and manual. The self-propelled type is friction driven by the power unit operating the saw, it runs on pneumatic tyres and is designed to hold a straight course without the aid of an external guide. The manually propelled type machine is guided by rails which must be set along the line of the joint in advance. Whereas there are obvious disadvantages to this system, manual control, by enabling the operator to assess the work being done by the saw, can result in economies in blade wear.

(7) Longitudinal joints may be constructed in the same manner as " dummy " joints, a strip of timber, concrete or joint filler being set on the formation to reduce the depth of

slot required at the surface. In order to minimize blade wear the joints should be cut in a straight line.

Buildings

The only type of joint normally required in buildings is the expansion joint the construction of which is relatively simple and has already been covered in the descriptions of the design of this type of joint for the various parts of a building. The chief consideration in the construction of joints in buildings is usually the production of a neat appearance. It is very important therefore that formwork is accurately constructed and that the concrete mix is sufficiently workable to permit thorough compaction under conditions which may be rendered very difficult by the amount and distribution of reinforcement. The use of a simple shape for the surface sealing cavity is particularly desirable. This cavity should also be formed by a fillet constructed for the purpose rather than by the joint filler.

The aesthetic aspects of the design, construction and treatment of joints are considered in detail in Chapter Ten.

CHAPTER TEN

The Location and Sizes of Joints

When deciding upon the dimension and shape and position of either a construction or a functional joint the following aspects should be considered :—

(1) The stability of the structure.

(2) The functional requirements of the jointing materials.

(3) The type of material conforming to the requirements.

(4) The finished appearance of the joint and the structure as a whole.

In the interests of stability the joints in concrete structures of all types should be placed where shear forces will be low and it has been suggested[41] that construction joints in beams and slabs should be at the centre of span or within the middle third.

Both the location and the size of joints will, in general however, be chosen according to the type of structure not only from the point of view of performance but also that of appearance.

WATER-RETAINING STRUCTURES

The joints in a water-retaining structure may be located either as continuous breaks throughout the structure or the roof and floor may be isolated from the walls. In both methods joints at the corners of the walls should be avoided in order to promote structural stability and to facilitate the installation of the jointing materials. The joints in the roof and floor should not be staggered because frictional restraint between abutting slabs may result in sympathetic cracking.

Location. When locating the joints in water-retaining structures the position of ancillary works such as penstocks, inlet and outlet pipes should be considered to ensure that the sealing cavities will be readily accessible for the initial application of the sealing compound and subsequent maintenance.

The ability of the waterbar to conform to horizontal and vertical curves will often influence the choice of location for the joints. In

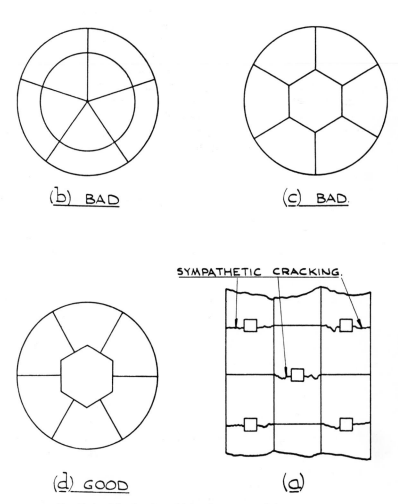

FIG 10. 1.   The location of joints in water-retaining structures.

general the installation of curved horizontal joints in floors and roofs should be avoided where possible. In circular tanks the walls may be constructed to rest on the floor using a sliding joint spanned by a waterbar provided with a degree of freedom at the interface. When a concentric joint in the floor slab is considered necessary it should be constructed in the form of a plane-sided figure to avoid horizontal curves in the waterbar. A five or six-sided figure is usually most appropriate in avoiding acute angles, radial joints being located at the centre of each side leaving the centre of the floor monolithic as shown in Fig. 10. 1. (d). Continuation of radial joints to the centre of the floor as shown in (b) leads to weakness of individual slabs and difficulties are usually experienced in placing and compacting the concrete in the acute angles.

Size. Expansion joints, when provided, can and should if possible be formed by installing a preformed joint filler ½ inch thick during construction. In exposed structures such as sea walls it may occasionally be considered desirable to provide an expansion space of ¾ inch but it is preferable to limit the length of the slabs to the values indicated in Table 2. 1 so as to reduce the need for wide joints. Narrow joints improve the appearance and minimize abrasion effects.

In deciding upon the width of expansion joints the size and shape of the sealing cavity must be given careful consideration. Sealing compounds tend to flow when exposed to sustained pressure such as that due to impounded water. It is most desirable, therefore, that the shape of the sealing cavity be such as to afford some support to the sealing compound. A few typical and simple designs are shown in Fig. 10. 2. Rectangular cavities are suitable for sealing compounds which are applied hot by pouring or by trowel. Strip form sealing compounds are most easily applied to triangular cavities the initial stage of installation being indicated by the dotted lines in Fig. 10. 2. (b) and (d). Triangular cavities permit easy removal of shuttering thus minimizing damage to the walls of the joint during construction. They also facilitate the placing and compaction of the concrete to obtain high density in this critical region. Both the triangular and the rectangular cavities illustrated, however, whilst providing a desirable increase in the width of the sealing compound to enhance its ability to accommodate movement, necessitate the adoption of a relatively thin joint filler. Hence the necessity of employing an expansion joint spacing which enables a joint width of ½ inch to be used.

There is no advantage in increasing the depth of a sealing cavity beyond 1½ to 2 inches. Not only is it extremely difficult to remove foreign matter and surface laitance from the walls of a deep cavity, but the following problems also arise :—

144

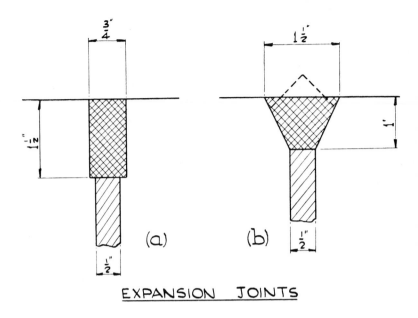

EXPANSION JOINTS

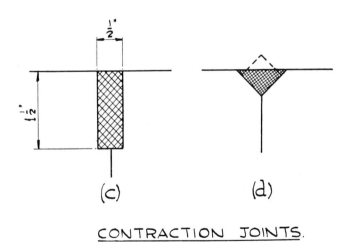

CONTRACTION JOINTS.

FIG. 10. 2. Sealing cavities for joints in water-retaining structures.

(a) The walls of the joint will not dry out readily.

(b) It is impossible to inspect the walls for cleanliness and freedom from dampness.

(c) It is impossible to apply primer satisfactorily.

(d) It is impossible to ensure that the primer has dried out thoroughly before the sealing compound is applied.

(e) It is difficult, without the danger of overheating, to ensure that the sealing compound is sufficiently hot on reaching the base of the cavity to develop proper adhesion to the concrete.

These criticisms of deep sealing cavities apply not only to water-retaining structures but to all other structures where the concrete is cast in situ. A further consideration is the " shape factor " to be discussed later in this chapter.

PAVINGS

There is little choice possible in the location of the joints in pavings but it is desirable that the following precautions be observed :—

(a) No joint should be provided between a pre-cast kerb and the main slab. Kerbs should be laid on the main slab and expansion and contraction joints should be located in the kerbs in line with the joints in the main slab. By this means the difficulties of providing a satisfactory seal under the difficult conditions existing at the edge of the slab are avoided. When placed on the slab the kerbs prevent traffic loading on the weakest part of the main slabs.

(b) Expansion joints should always be provided at the junction of a paving and another type of structure such as a building or a bridge. They are also necessary where two pavings intersect.

(c) Staggered joints are undesirable because of the tendency for sympathetic cracking. The use of continuous reinforcement, tie-bars or dowels will inevitably cause cracking of this type and load transfer devices, of which these are the most efficient, are almost invariably necessary to preserve a smooth running surface in a paving.

(d) Joints should not be installed close to services manholes or gullies because this further weakens the end of a slab which is already the weakest part.

Size. The sizes of the joints in a concrete paving will vary according to the design of the paving and the method of forming the joints. It has already been suggested that in view of the great reductions in surface irregularity made possible by the technique of sawing joints, the installation of joints at fairly close spacings may be an advantage. If this design is adopted and the spacing of expansion joints is reduced to 30-40 ft., these joints may be constructed using a $\frac{3}{8}$ inch thick joint filler. When the slab length is greater than 60 ft. however, $\frac{3}{4}$ inch or 1 inch thick filler should be used. Wide expansion joints are particularly necessary when intermediate dummy joints are installed or cut because some of these joints will open progressively and reduce the expansion space available to accommodate movement due to increases in temperature.

The dimensions of the sealing cavity in expansion joints should be chosen in such a way as to minimize extrusion of the sealing compound in hot weather. Deep, narrow cavities should not be used. Consider, for example, an expansion joint in a paving having a joint spacing of 100 feet with no intermediate " dummy " or contraction joints. Assume that the joint width is reduced by $\frac{1}{8}$ inch during a spell of hot weather. If the sealing cavity is $\frac{1}{2}$ inch wide and $1\frac{1}{2}$ inch deep the volume of sealing compound per inch run is $\frac{3}{4}$ in.^{3} The reduction in width due to expansion will reduce the volume of the sealing cavity by $\frac{3}{16}$ inch^3. The volume of the sealing compound however is constant so that when the joint is $\frac{3}{8}$ inch wide displacement of the material above the surface of the concrete will result in a hump not less than $\frac{3}{16}$ to $\frac{3}{8}$ inch in height.

This " shape factor " is also important in its effect on both the bond and the cohesive strengths of the sealing compound when the joint opens. Calculated values for the strain in the joint sealing compound due to 100% extension showing the influence of the width/depth ratio are illustrated in Fig. 10. 3.

Thus, depending to some extent on the movement to be accommodated, it would appear that the depth of the seal should never exceed the width. For most purposes and particularly where large movements are anticipated, the depth should be less than the width, with the overriding proviso that the depth should not be less than $\frac{1}{2}$ inch. Shallower depths than this do not provide sufficient latitude for preparation of the joint and application of the sealing compound. In water-retaining structures where expansion and contraction of the concrete is minimal, sealing compound depths of up to $1\frac{1}{2}$ inches are permissible and in fact desirable to prevent water under pressure from by-passing the seal through the concrete.

BUILDINGS

The most important consideration governing the location and size of the joints in a building is usually the appearance of the structure as a whole. Wherever it is possible to do so, joints should be incorporated as an architectural feature. Construction joints can be masked by grooves, a line of tiles or a similar surface finish, but expansion joints require more than this because

(a) they are designed to accommodate movement, and

(b) they must be sealed to prevent water passing through them.

The location of expansion joints.   The best location for an expansion joint in a building is along a natural line provided by doors or windows.

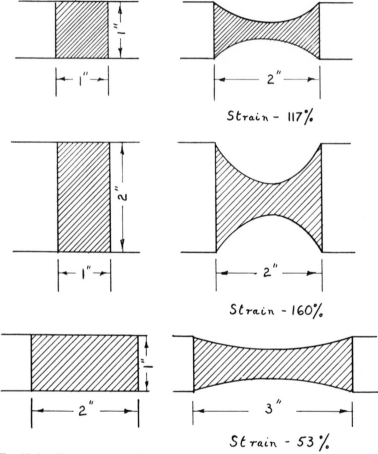

Strain - 117%

Strain - 160%

Strain - 53%

Fig. 10. 3.   The influence of width/depth ratio on the strain on the sealing compound at 1 inch extension.

A typical example of such a joint incorporating a white P.V.C. insert is shown in Fig. 10. 4. Not only does this treatment reduce the length and prominence of the joint, but being in a reduced section of the structure, the joint is well placed to prevent the development of cracks due to contraction

FIG.10.4.   A white P.V.C. insert in an expansion joint in a building.

When a building is constructed in wings, the expansion joints can be installed at the junctions in the structure and they are then hidden by being in corners.   Full use should be made of corners in the location of expansion joints in a building.   The expansion joints in floors are best located at or near the junction between the walls and the floor thus avoiding, except in very large floors, wide joints across the main floor area.   A novel architectural effect which makes use of the joints as a feature in a garage forecourt is shown in Fig. 10. 5. Similar treatments could be applied to the floors and even to the walls of buildings when the use of a black sealing compound is obligatory from the point of view of performance.

Joint widths. In order to minimize the number of joints required in a building it is frequently necessary to provide an expansion space of $\frac{3}{4}$ to 1 inch at each joint. When designing a building on this basis however the method of sealing the joints in the walls must be given

FIG. 10.5. The use of a pattern to conceal joints in a garage forecourt.

due consideration. Wide joints in walls can be sealed by means of three types of material as follows :—

    (i) A copper strip which is set into the concrete during construction.

    (ii) A bituminous vertical joint sealing compound.

    (iii) A building mastic used in conjunction with a P.V.C. insert.

Polysulphide and other synthetic rubbers might be another alternative but insufficient experience is available on the use of these materials to enable definite recommendations on their use to be made here.

Of the three conventional sealing methods only the last permits a choice of colour and in using this method it must be remembered that P.V.C. inserts must be maintained in a state of compression and therefore will accommodate only limited contraction. The other two methods should be incorporated as architectural features or be suitably concealed by being located in corners.

If a maximum spacing of the order of 60 ft. between expansion joints is adopted a joint width of $\frac{1}{2}$ inch is normally sufficient. Joints of this width can be effectively sealed with cold-applied building mastics and in this case a wide range of colours is available. Thus although joints at close spacings may present some measure of inconvenience during construction they may be less obtrusive than the wider joints made necessary by larger spacings and in addition they will be more likely to give satisfactory performance.

It has already been indicated that larger movements may occur in roofs than elsewhere and to reduce the strain on the sealing material the joints or the sealing cavities in roofs should preferably be not less than one inch wide.

The widths of the joints in floors should always be restricted to the minimum considered necessary to prevent the development of deleterious stresses in the concrete. Wide joints in floors lead to spalling of the edges of the slabs and increase the possibility of the sealing compound being displaced by traffic. However, it is usually necessary to instal shrinkage joints at fairly close spacings in floors, so that expansion joints may not be necessary except to isolate the floor from the walls. In this case traffic effects do not arise. When considered necessary expansion joints should not be more than $\frac{1}{2}$ inch wide and the edges of the slabs should be protected by a P.V.C. insert having a convex upper surface, a key-way being provided in the side of the insert to retain it in the joint. Similar inserts $\frac{3}{16}$ or $\frac{1}{4}$ inch thick may be used in shrinkage joints. P.V.C. inserts protect the edges of the slabs and prevent the ingress of foreign matter ; they do not prevent water entering the joint and if a seal against water penetration is necessary a sealing compound should be applied to the joint before the P.V.C. insert is installed. This will usually involve the formation of a cavity in the concrete base slab to accommodate the sealing compound, the insert then being installed in the wearing course.

Obviously there are many architectural considerations which will affect the design, location and sizes of the joints in buildings. Whatever these considerations may be, however, they must be examined in juxtaposition to the problem of sealing the joints and the limitations of the materials available for doing this must not be overlooked. The treatment of defective joints in buildings is often very costly and if severe conditions are imposed on the sealing materials by the general design of the structure there may be no form of treatment which will effect a substantial improvement in performance.

CHAPTER ELEVEN

Specialized Joint Design

The general principle of providing joints to prevent the development of tensile stresses can usually be applied with advantage to concrete structures of all types. There are however a number of structures and concrete members which present particular problems in the selection of suitable designs and jointing materials. A few of these special cases will now be considered in some detail.

BRIDGES

The jointing technique applied to a concrete bridge requires particular care not only to ensure that structural stability is maintained in a structure whose performance cannot readily be assessed, but also to preserve its beauty. The conditions to be accommodated are often difficult to predetermine but there are a few common effects for which provision may be made in the design of various sections of the structure.

Abutments. Movement may tend to occur in abutments due to earth pressure, differential shrinkage, temperature variations and to settlement. In order to prevent the development of cracks where these effects cause excessive tensile stresses, joints should be provided unless a rigid frame structure is used. When joints are installed abutments may be considered in four sections as follows :—

(1) The footings. Footings may move both horizontally and vertically due to pressure of the backfill, irregular foundation conditions and the different loading conditions imposed by the wing walls and the breast wall. Sagging at the centre of the breast wall is a common tendency particularly when the footings are continuous beneath the breast wall and the wing walls. In order to distribute the foundation pressure uniformly, construction joints at the top of footings may be designed as hinge joints similar to that shown in Fig. 11. 1. As a further precaution the amount of reinforcement at the centre of the footings may be increased.

152

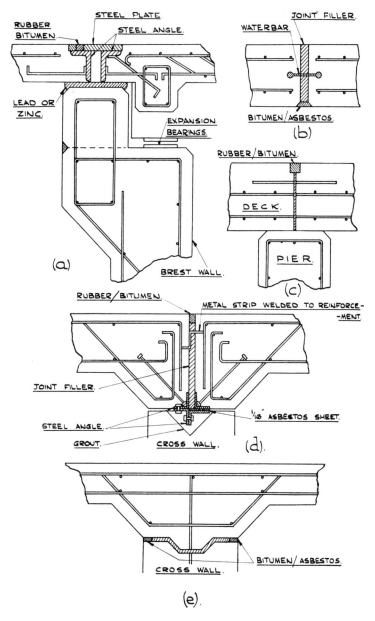

FIG. 11. 1.   Various types of joint used in bridges.

(2) The wing walls. The loading conditions on wing walls are significantly different from those on the breast wall. Flared and parallel wing walls may in practice act as counterforts with the development of high tensile stresses. The introduction of extra reinforcement near the top of the walls will minimize the effect of these stresses but the provision of joints isolating the wing walls from the breast wall is probably the most satisfactory solution. A typical joint of this type is shown in Fig. 11. 2. It is desirable that a waterbar be incorporated in these joints in order to prevent disfigurement due to seepage of water from the back fill. Provision should, however, be made for the back fill to drain because increases in the water content of the back fill will produce heavier loading conditions on the wing walls and the breast wall. Dense back fill may be drained by weep holes with a granular backing as shown in Fig. 11. 3. The use of horizontal grooves to conceal lift joints is also shown. These joints are potential planes of weakness and should be provided with additional tie-bars together with a waterproofing medium on the earth face. The staining in this example was due to contamination of water draining from the surface through the joint prior to the application of a sealing compound during the construction period.

In cast-in-situ abutments an expansion joint such as (b) in Fig. 11. 1. is more suitable than an open joint. This joint contains, in addition to the waterbar, a compressible joint filler which is protected at the surface by a bituminous vertical joint sealing compound.

(3) The breast walls. Expansion joints are not normally necessary in breast walls less than fifty feet long. In order to prevent the development of irregular shrinkage cracks, however, contraction joints should be installed at 15-20 ft. spacing. These joints should be adequately sealed on the earth face either by a sealing compound in a groove, dowel bars being set across the joint, or by a waterproof membrane. Grooves should be formed in the exposed surface to conceal these joints.

(4) The bridge seating. A typical arrangement for providing allowance for expansion of the bridge deck and approach slab at the bridge seating is shown in Fig. 11. 1. (a). In this type of construction it is important that the expansion bearings be kept clean and dry to prevent binding and thus effective seals must be provided at all joints. In addition,

FIG. 11. 2. An expansion joint in the wing wall of a bridge.

therefore, to the seal in the running surface a membrane or sealing compound should be applied to the construction joint in the breast wall and a corrosion resistant metal plate which will also permit freedom of movement should be installed between the back wall and the running slabs.

The provision of seals at the running surface and at the construction joint in the breast wall is equally important at fixed bearings in order to prevent corrosion of the reinforcement and tie rods which pass through these joints.

In multi-span bridges it is usually necessary to construct joints to relieve horizontal thrust due to expansion and to rocking. Construction joints are a source of weakness and frequently crack due to shrinkage and other stresses. They should therefore be avoided as far as possible by constructing deliberate joints such as that shown in (c) Fig. 11. 1. to allow for movement. The essential features of this design are as follows :—

(i) Additional tie-rods are provided near the top of the slabs to restrain the tendency for the joint to open.

(ii) A groove is formed at the surface for the application of a hot-poured sealing compound which will prevent water passing through the joint.

Alternatively the reinforcement may be stopped short of the joint and the deck slabs tied by dowels to the pier. In this case a thin strip, say ¼ inch, of joint filler should be provided between the abutting slabs to accommodate deflection.

Expansion bearings similar to that illustrated for the deck slab in (a) Fig. 11. 1. may be used at appropriate intervals in order to reduce horizontal thrust. The number of these joints should be reduced to a minimum by the use of rocker bearings and flexible columns. A typical arrangement of rocker and expansion bearings used in a multi-arch bridge is shown in Fig. 11. 4. The rocker bearings (type (e) in Fig. 11. 1.) permit no horizontal movement and therefore alternate with expansion joints, those denoted by (d) permitting only one section to slide as shown in Fig. 11. 1. whereas those at (f) should permit both sections to slide, an asbestos or neoprene membrane being provided underneath both sections.

When expansion space is provided in bridge decks the joint should be continued through the parapet walls. These joints may either be left open or they may contain a conventional non-extruding joint filler sealed at the surface with a vertical joint sealing compound. The joints are most discreetly located at the junction with pillars.

In skew bridges precautions should be taken to prevent lateral creep at expansion bearings. One possible method of construction

FIG. 11. 3.   A weep hole in the wing wall of a bridge.

to avoid this effect is to step the central portion of the bridge seating at these bearings so as to form a retaining block between the girders.

Only the simple guiding principles relating to the provision of joints in bridges have been considered but it is often found that simple treatments are the most effective for general use. In applying these principles the following suggestions[42] on the lay-out of bearings may be of some assistance :—

(1) In single-span lay-outs use two fixed bearings on the abutments when the span length does not exceed 45 ft.

(2) In two-span lay-outs use fixed bearings on the abutments and expansion bearings on the pier.

(3) In three-span lay-outs use fixed bearings on the abutments, two fixed bearings on one pier and two expansion bearings on the other.

(4) In four-span lay-outs use fixed bearings on the abutments and on the centre pier and use expansion bearings on the other two piers.

(5) In addition to transverse joints a longitudinal joint is desirable in a dual-carriageway deck.

MASONRY CONSTRUCTION

It is common practice to install expansion or contraction joints where masonry or brick wall panels meet or span reinforced concrete columns. These joints assist in preventing the development of high tensile stresses in the panels but they may not reduce the stresses sufficiently to eliminate cracking, particularly at reduced sections such as window and door openings. The tensile stresses which lead to failure are due to the combined effects of shrinkage and the edge restraint offered by foundations, floors and roof. Shrinkage of the masonry units after erection can be considerably reduced by drying processes after manufacture and the provision of suitable dry storage conditions on the site. These precautions however have been shown[43] to account for approximately 50 per cent of the total volume changes occurring in precast masonry blocks which, in addition to exhibiting initial shrinkage, will also undergo dimensional changes due to the variations between internal and external exposure conditions. The use of reinforcement in horizontal joints and of two or three layer systems is of some assistance in reducing these changes but the only completely satisfactory solution is the provision of contraction joints.

Joint spacing. There are so many factors which influence the tendency for cracks to develop that it is extremely difficult to assess how many joints will be required in a structure to ensure complete

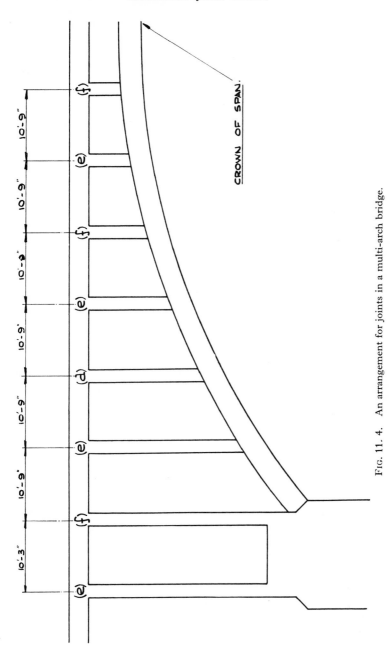

FIG. 11. 4.   An arrangement for joints in a multi-arch bridge.

freedom from cracking. Relationships are being developed[44] in terms of the length and height of walls and these provide a general guide to the spacing at which joints should be installed. A simple equation for walls highly restrained at top and bottom edges based on the strengths of the materials is as follows :—

$$L = 2 \left(\frac{f_t}{f_v} \right) . \left(\frac{A_t}{A_v} \right) H.$$

where L = Length of the wall in feet.

f_t = Average tensile strength of the block in lb./in.²

f_v = Average shear strength of the mortar joints in lb./in.²

A_t = The area of wall effective in resisting tensile stress per foot of wall height in sq. inches.

A_v = Area of mortar joint effective in resisting shear per foot of wall length in sq. inches.

H = Height of wall in feet.

This relationship does not allow for variations in the properties of the masonry blocks but in the absence of a fully developed approach embodying these additional variables, it could conceivably be used successfully by applying an appropriate factor of safety.

Typical values for the terms in the equation are as follows :—

$\dfrac{A_t}{A_v}$ = 0.5 for hollow concrete block construction.

f_t = 10 per cent of the crushing strength of the concrete i.e., 150-250 lb./in.², say 200 lb./in.²

f_v = 90-100 lb./in.² say, 100 lb./in.²

Substituting these terms in the equation, the relationship becomes $L = 2 H$ and thus, allowing a factor of safety of 2, joints should be installed at a spacing equal to the height of the wall.

A more fundamental approach[44] relating the length of wall, i.e., the joint spacing, and the height is based on the stress conditions at the centre line of the wall as determined by experiment. This relationship can be extended by assessing the stress conditions which may be anticipated due to shrinkage, temperature variations and the degree of restraint in a particular project but as work is still in progress on this approach the reader is referred to the paper published on the subject.

From practical observations on the performance of buildings it would appear to be desirable to provide contraction joints at spacings of not more than 30 feet in walls of uniform section and at a somewhat closer spacing where window and door openings cause stress concentrations at reduced sections. Corners should be isolated from adjoining walls maintaining the joint spacing used in straight sections.

Types of joint. In addition to the expansion joints which are necessary in the reinforced concrete frame, contraction joints should be provided in the masonry cladding. If the masonry spans the columns as an outer skin the joint may be continued through the cladding. A neat example of the latter method of construction using brick and masonry cladding is shown in Fig. 11. 5. Here the staggered joint in the brickwork is sealed with a light-coloured building mastic and the straight joints between the masonry blocks are filled with a P.V.C. insert bedded onto a mastic in the base of the joint.

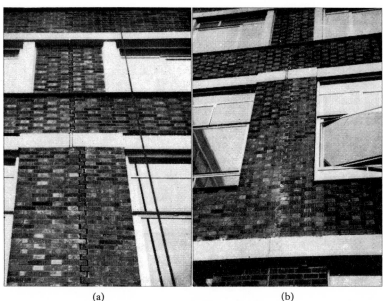

(a)   (b)

Fig. 11. 5.   A staggered joint in brick cladding, (a) before sealing ;   (b) filled with a light-coloured building mastic.

When the cladding butts against columns, joints should be provided on each side of the column as shown in Fig. 11. 6. (a). Other designs for joints in hollow block construction are shown in (b), (c), (d) and (e). The functions of the joints are as follows :—

(b) is a laterally stable joint designed to relieve stress concentrations at reduced sections.

(c) is designed for isolating intersecting load-bearing walls where lateral stability is required.

(d) provides no lateral stability and may be used to isolate a partition wall carrying no load.

(e) relieves the stress at the top of the foundations.

161

L

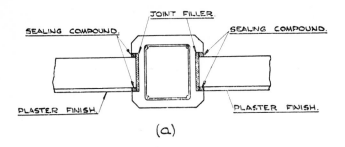

SEALING COMPOUND. JOINT FILLER. SEALING COMPOUND.

PLASTER FINISH. PLASTER FINISH.

(a)

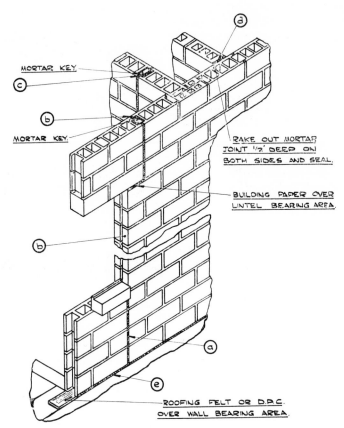

MORTAR KEY.

MORTAR KEY.

RAKE OUT MORTAR
JOINT 1/2" DEEP ON
BOTH SIDES AND SEAL.

BUILDING PAPER OVER
LINTEL BEARING AREA.

ROOFING FELT OR D.P.C.
OVER WALL BEARING AREA.

FIG. 11. 6. The location of joints in hollow block construction.

These joints are typical of the methods which may be applied to masonry and brick infilling panels in order to relieve stresses due to shrinkage and temperature variations and differentials. Clearly they do not embrace all the design possibilities but they do cover the most important aspects to be considered if random cracking is to be avoided.

Jointing materials. When the joints are suitably concealed, rubber bitumen or bitumen/asbestos sealing compounds are efficient and economical due to their good weathering properties. The most commonly used compounds however are resin-based building mastics and polymeric compounds. In addition to being available in various colours these compounds have the advantage of being applied by means of a gun and therefore can give a neat appearance. P.V.C. cover strips are also suitable but must be supplemented with a sealing compound in external joints to ensure weather protection.

Joint fillers, when required, must be non-extruding and in hollow glass block panels should be readily compressible ; felted mineral fibre is particularly suitable. The usual precaution of applying a sealing membrane must be observed when a resinous building mastic is used to seal a joint filled with an impregnated joint filler.

PREFABRICATED CONSTRUCTION

The light framework and cladding used in prefabricated construction methods of building, commonly known as curtain walling can result in very large movements due not only to moisture and temperature variations, but also to wind pressure and to the heterogeneous nature of the structure. It is common practice therefore to provide flexible joints between the units both horizontally and vertically.

The most successful method of providing flexible joints between prefabricated units is probably by means of the preformed strip sealing compounds. These materials develop adhesion on contact and can therefore be set on horizontal courses and vertical stanchions in readiness for the prefabricated units being erected against them. Typical methods of using these materials in vertical and horizontal joints are shown in Fig. 11. 7. In the vertical joint (a) the strip sealer is set onto the stanchion and the unit is clamped against it, a joint filling material being interposed between adjacent units to provide support for a surface sealing compound which can be a building mastic or one of the recently developed synthetic rubber compositions.

The horizontal joints (b) and (c) provide easy drainage and the strip sealer, whilst acting as a bedding material to accommodate discrepancies in the dimensions of the units, is contained in cavities which

163

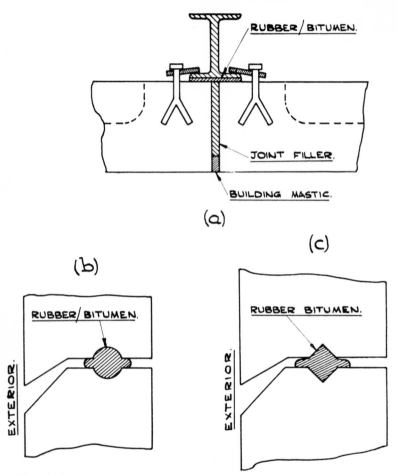

FIG. 11. 7. The use of strip form sealing compounds in masonry cladding.

ensure adequate initial deformation of the sealer under the loading of the units.

It is evident in the light of experience that particular care must be paid to the sealing of the joints in prefabricated construction. When a building mastic is used it should have good oil retention, low solvent content and high resistance to oxidation and whenever possible a relatively inert material such as bitumen or a rubber bitumen composition should be installed as a second barrier to the ingress of water.

The most recent approach to the sealing of joints in prefabricated construction, more particularly in association with high-rise buildings, is the use of " drained " points. This system is an extension of the designs shown in (b) and (c) of Fig. 11. 7. The joints between abutting panels in the cladding of the structure are designed to permit free drainage and a diaphragm baffle plate is set to span the joints between the panels well below the surface. Numerous designs of baffle plates and cavity dimensions with and without bonded sealing compounds at the back of the joints as shown in (a) Fig. 11. 7. have been used[44]. Indeed, it would appear that the use of drained open joints is the most promising method of preventing the penetration of rain when large concrete cladding panels are used in high-rise buildings where high run-off rates result in relatively high dynamic water pressures on surface sealing compounds.

PRESTRESSED CONCRETE

The application of a compressive stress to a concrete member in order to reduce the tendency for the development of tensile zones to a large extent overcomes the need for the introduction of joints. It is frequently necessary, however, to install joints during construction to accommodate the movement which occurs during the stressing process. In many structures these joints will only be temporary and sealing devices are installed merely to assist in retaining the cement grout used to bond sections together after stressing. In other structures however, permanent joints are necessary and it is desirable therefore to consider the circumstances under which these two design methods have been used.

Bridges. Rotation and expansion is often accommodated in prestressed concrete bridges by precast reinforced concrete hinges which have reduced sections or throats $\frac{5}{8}$ to $\frac{3}{4}$ inch thick to create high stress concentrations in the concrete. Under the high stresses which develop at these throats plastic flow occurs which, together with the strain induced in the reinforcement, accommodates the movement. In some cases however hinges and bearings similar to those used in reinforced concrete bridges have been employed. Three designs which have been used in bridges constructed by the Freyssinet system are shown in Fig. 11. 8.

The use of rubber bearings as shown in (c) is considered[45] to be a simple method of providing for rotation and expansion in cantilever construction and bearings of this type are applicable to wide and to skew bridges. In designing these joints however the distance x must be large enough to permit adequate distribution of the prestress in the bearing cross-section and y must not be large enough to allow the flexural resistance to be exceeded.

Rubber bearings comprise alternate layers of rubber sheet and perforated sheet to prevent permanent deformation under load. In long

165

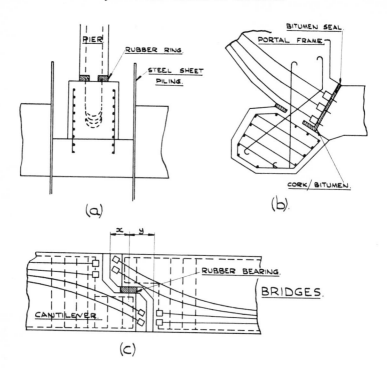

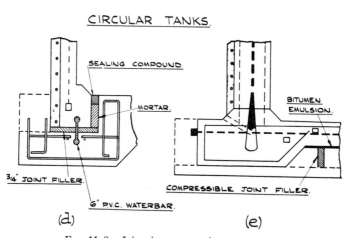

FIG. 11.8.   Joints in prestressed concrete structures.

spans the thickness of the bearing differs at the two ends, a thin layer behaving as a fixed bearing and a thick layer as an expansion bearing.

Sliding plate and roller type metallic bearings can also be used but these are more difficult to install than rubber bearings which are, as a result, becoming more widely used.

Pavings. The most important limiting factor in the construction of prestressed concrete pavings is the restraint offered by the subgrade to free movement of the slabs. This restraint reduces the prestress and in extreme cases may cause tensile stresses in the concrete. An efficient sliding layer should therefore be provided to isolate the slabs from the subgrade. A uniform layer of sand ⅛ inch thick covered with waterproof paper has been found[46] effective in reducing the value of subgrade restraint but care must be exercised to ensure that the sand layer is loose and is screeded to a smooth finish.

Another method of reducing subgrade restraint which has been used in a paving constructed in precast vacuum concrete slabs comprises the erection of dwarf walls on the prepared foundation prior to placing the slabs in position. After the prestress had been applied, the cavities beneath the slabs were grouted.

A value of 1.5 has been suggested[47] for the coefficient of friction due to subgrade restraint and this limits the maximum practicable length of slab to about 400 feet. Research into methods of reducing this coefficient appears to merit consideration so that the construction of longer slabs can be achieved. Research is also required into methods of filling and sealing the joints between these very long slabs. Clearly it is undesirable to install joints wider than about 1½ inches in the interest of riding quality and as movements between adjacent 400 ft. slabs may amount to ¾ inch or more, filling and sealing materials of very high performance are called for. Foamed plastics and polysulphide rubber compositions may provide a satisfactory solution but much experience has yet to be gained in the formulation and installation of these materials.

Water-retaining structures. The use of prestressed concrete is particularly appropriate to the construction of circular water tanks. As in the case of pavings, the reduction of restraint to the deflection of the concrete during the stressing operation demands the provision of sliding joints in medium or large diameter tanks. The amount of movement occurring at these joints is unlikely to be very large. Measurements taken during the construction of a 75 ft. diameter tank having a sliding joint at the foot of the wall indicated[48] a total average inward movement of the wall of 0.18 inch across its diameter. The movement to be accommodated at the joint therefore was less than 1/10th inch.

In addition to the usual precautions which should be taken to

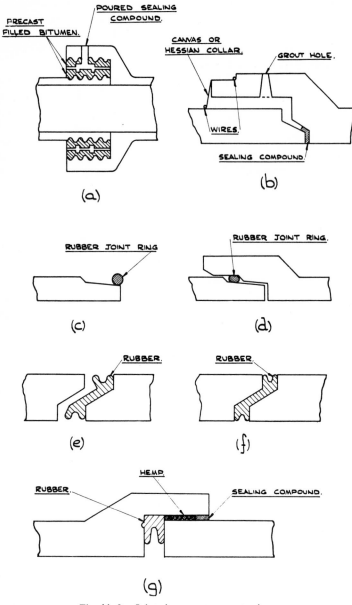

Fig. 11. 9.   Joints in precast concrete pipes.

obtain a smooth bearing surface at the sliding joint it has been found that the installation of a flexible P.V.C. or rubber waterbar is of considerable assistance in ensuring that the joint is liquid-tight. Failure to incorporate a waterbar has resulted in leakage as a result of movement caused by variations in temperature. When the joints are grouted after stressing the waterbar also provides an effective seal to prevent loss of grout.

Typical examples of joints which have been used successfully are shown in Fig. 11. 8. (d) and (e). In (d) prestress is applied to the wall only and to permit the waterbar to distort the wall rests on a compressible joint filler. This design is particularly appropriate to large diameter tanks and was used in a 105 ft. diameter open-topped tank designed to hold oily water at a temperature of 80°F. (It is interesting to note here that polysulphide synthetic rubbers have been used to line concrete oil storage tanks in order to prevent deterioration of the concrete).

Smaller tanks can conveniently and economically be constructed monolithically as shown in 11. 8. (e), a compressible joint being formed to isolate the sub-base from the foundation ring beam.

CONCRETE PIPES

Although concrete pipes are usually protected from temperature variations by being buried, it is frequently necessary to provide flexible joints in order to accommodate deflections caused by soil movements. Moreover in many large hydro-electric projects, pipelines are often laid above ground and are then exposed to the very severe conditions experienced in hilly or mountainous terrains. It is axiomatic that the joints in pipelines must be efficiently sealed and this raises a number of problems as follows :—

(1) The sealing medium must be applied either before the pipes are brought together or from the outside after placing.

(2) Water pressure is exerted against the inner surface of the pipes and will displace conventional bituminous sealing compounds unless they are given a rigid backing.

(3) Reliance is often placed on abutting surfaces of adjacent pipes for accurate alignment so that a flexible seal on the inside is undesirable.

(4) Very low temperatures may be experienced which will result in considerable contraction of the concrete and will also cause most bituminous materials to become brittle.

(5) The joints are in a vertical or near vertical plane so that conventional hot-poured sealing compounds will tend to flow out of the joints both during and subsequent to appli-

cation if not supported.

A number of typical joint designs which have been developed to overcome some of these problems are shown in Fig. 11. 9.

Design (a) has been developed for salt-glazed earthenware pipes but is included here because it could conceivably be used in small diameter concrete pipes. As shown in the sketch the space between the spigot and socket is occupied by a filled bitumen cast onto the pipes before leaving the manufacturer, a small channel being formed at the centre for subsequent filling with a hot-poured sealing compound Two holes are formed in the wall of the socket, one to provide an entry for the sealing compound and the other to allow air to escape and to indicate when the joint is full. The application of the sealing compound on site must be carried out as expeditiously as possible to ensure that cooling does not take place before the joint has been filled. When correctly applied, the hot sealing compound should blend with the pre-formed composition in the vicinity of the channel, elsewhere however the high filler content of the preformed material causes it to have high resistance to flow under water pressure.

In design (b) a thin layer of soft sealing compound is set into the socket before the pipes are placed and the remainder of the cavity is filled with mortar grout which is contained by a collar of hessian or canvas tightly wired onto the two pipes. This joint should not be expected to accommodate more than relatively small longitudinal movements.

Three designs which give a high degree of flexibility incorporate preformed rubber rings which are set onto the spigot or into the socket before the pipes are placed. The rubber is then compressed by jacking the pipes together as shown in (c) and (d) and in (e) and (f). Two refinements may be applied to (d) as follows :—

(i) Means may be provided for a poured sealing compound to be applied between the rubber ring and the water face.

(ii) Simple butt-jointed pipes may be used with a rubber ring on each, held in position by a rebate in a collar which is forced over them to span the joint.

It seems unlikely that any of these designs completely fulfills the requirements of a pipeline which must accommodate large movements. This is a problem which has yet to be solved and is, in fact, another possible application for polysulphide rubber/resin compositions. At the present time, however, the high cost of these materials may preclude their use in the lage cavities provided in the larger diameter pipes which present the worst conditions.

CHAPTER TWELVE

The Application of Sealing Compounds

Before considering the practical aspects of preparing and sealing joints it may be as well to summarize the general functions and properties of sealing compounds as discussed in Chapter VII as follows :—

(1) Adhesion. The development of a strong adhesive bond to the walls of the joint is always essential. Although the strength of the bond depends to some extent on the characteristics of the sealing compound, the performance of all sealing compounds is greatly influenced by the condition of the joint and also the condition of the material when it is applied.

(2) Extensibility and resistance to flow and the ingress of foreign matter. The degree to which these properties are required in a sealing compound, controls its temperature susceptibility. When the properties are required to a high degree, e.g. in the case of compounds for sealing vertical joints exposed to severe weather conditions, a very low temperature susceptibility is necessary. In hot applied compounds this can, and usually does, call for careful control over the heating process to ensure that the compounds are sufficiently workable and that their properties are not affected adversely in rendering and maintaining them in this condition. Moreover even under the most favourable conditions it is likely that much diligence will be necessary in applying the sealing compound to ensure that the joint is properly filled and therefore impermeable.

WHEN TO SEAL

The most appropriate time for applying the sealing compound is usually chosen to suit the general construction programme. Whereas this approach may be favoured by the economic execution of the work, however, it may mean that the sealing operation is carried out under conditions which are not conducive to successful subsequent performance.

Probably the most important single factor affecting the performance of sealing compounds is dampness. Although development work now in progress indicates that certain synthetic rubber/bitumen and tar compositions will adhere to damp concrete, no methods have yet been evolved which effect any significant improvement in the adhesion of conventional sealing compounds to damp concrete. It is highly desirable therefore that the sealing operation be carried out in dry conditions. Structures such as covered reservoirs are best sealed before the roof is constructed so as to avoid difficulties arising from condensation. The floors of buildings should be sealed as near to the completion of the project as possible. Joints exposed to the elements should be sealed during a period of dry weather.

Whenever possible consideration should be given to the subsequent movement of the concrete. Whereas dry summer conditions would appear to be particularly favourable, sealing at the warmest time of the year means that subsequent contraction of the concrete will be very large and thus will impose considerable stress on the adhesive bond of the sealing compound. The most appropriate time for sealing the joints in exposed structures is during dry spells of weather in the Spring. The annual movement is then more or less equally divided between expansion and contraction so that a compromise is made between extrusion and the development of high bond stress in the sealing compound.

CLEANING THE JOINT

In order to provide every opportunity for the concrete to dry, the sealing cavity and the walls of the joints should be cleaned and exposed to the atmosphere as soon as possible after the construction of the joint.

Formed cavities. The sealing cavities in horizontal joints in pavings and floors which are formed by the joint filler should be raked out to the appropriate depth ($1\frac{1}{2}$ inches in water-retaining structures and up to 1 inch in other structures) either by hand methods or using a rotary cutter of the type shown in Fig. 12. 1. This is a modification of the rotary wire brush shown in operation in Fig. 12. 2. and comprises a 4.5 B.H.P. petrol engine driving a double knot circular wire brush at approximately 2,000 r.p.m. through a treble V-belt coupling. The depth of penetration into the sealing cavity is controlled by the plumber block at the front of the machine. When used for removing joint filler various types of blade have been used

FIG. 12. 1. A rotary cutter for removing joint filler from the tops of expansion joints in pavings.

Photograph by permission of Expandite Limited, London

and it appears that the most economical and effective design[49] incorporates two or four hardened steel removable tines which are rivetted or dowelled into two central boss plates.

Having removed the greater part of the joint filler from the top of the joint the walls of the cavity should then be thoroughly brushed to remove laitance, dust and fibres from the surface. No equipment is available for the treatment of vertical joints and these must therefore be raked out and brushed by hand or manually operated abrasive blades or by blasting.

FIG. 12. 2. A rotary wire brush being used in the maintenance of joints.
Photograph by permission of Expandite Limited, London.

Sawn cavities. The use of abrasive blades for cutting joints in concrete pavings has led to a number of difficulties in cleaning the walls of the sealing cavity. In order to facilitate cleaning and the subsequent application of sealing compound it is recommended that sawn grooves be made not less than $\frac{1}{4}$ inch wide or more than $\frac{1}{2}$ inch deep. No method which is considered completely satisfactory has yet been developed for cleaning these cavities and probably the most effective treatment at the present time is to wash the joints with a jet of water immediately after the passage of the concrete saw and then follow up with a high pressure jet of compressed air or by blasting with sand or grit. It does not appear to be satisfactory to allow the grooves to dry between sawing and cleaning because the fine slurry produced during the sawing operation contains cement which has not fully hydrated and this bonds as a thin layer. Removal of this layer after it has set will obviously be difficult since it is not possible, due to the width of the grooves, to brush the walls effectively.

Subsequent to this initial cleaning of both formed and sawn sealing cavities, the concrete should be allowed to dry and as far as

FIG. 12. 3. A primer spraying unit which incorporates an air jet for the removal of loose dust.
Photograph by permission of Expandite Limited, London.

possible loose dust and foreign matter should be removed from the whole area to be sealed. When the concrete has dried the sealing cavities should again be brushed and loose matter removed by a jet of compressed air, a moisture baffle being incorporated between the compressor and the nozzle or preferably by blasting.

PRIMING

Laboratory work on the extensibility of sealing compounds at low temperatures has indicated that the application of a primer to the surface of the concrete can often improve performance.

The function of the primer. The exact function of a primer when used with hot applied sealing compounds has not been established but it is likely that it serves to coat any particles of dust in the joint so that they mix with the sealing compound when it is poured instead of forming a parting layer between the compound and the concrete. The coating of primer may also reduce the development of bubbles at the surface of the concrete caused by the expansion of air in the pores. Apart from reducing the area of contact between the sealing compound and the concrete these bubbles can also lead to other deleterious effects as will be described in Chapter XIII.

When used with mastics a primer fulfills two functions. It seals the surface of the concrete so as to inhibit the absorption of oil from the mastic and it also minimizes the risk of saponification which may occur with some types of mastic in alkaline conditions.

A primer also fulfills a particular function in the application of cold-applied rubber/bitumen and foamed plastic strip sealing compounds because in this case it behaves as an adhesive to hold the strip in position while the joint is being made.

Types of primer. The most effective type of primer for use with hot-applied bituminous sealing compounds appears to be a blend of high viscosity bitumen or tar and a low boiling point solvent. The proportion of the high boiling point fraction should not be more than about 30 per cent in order to promote penetration and to avoid the formation of a thick layer of tar or bitumen on the surface. A thick film of primer will prevent adhesion of the sealing compound by forming a parting layer between the sealing compound and the concrete. Small proportions of wetting agents may also be of value in combatting damp conditions ; such additives should not, however, be used in primers containing materials which will saponify. Another type of primer which shows promise is a polyamide cured tar oil/ epoxy composition which has the property of preferentially " wetting " damp surfaces and providing a good bond when the sealing compound is applied within 20 minutes of the application of the primer.

The simplest and generally most satisfactory primer for use with resinous building mastics is shellac or " knotting ". There are, however, several proprietary primers which have been developed specifically for sealing porous surfaces and as alkali sealers.

Method of application. The most efficient method of applying primer is undoubtedly by brush but this is a tedious operation if carried out correctly and there is a tendency for operators to use a full brush in order to cover the surface, instead of brushing out and stippling into the irregularities. This can result in an excess of primer being applied at intervals along the joints and this may lead to premature failure of adhesion between the sealing compound and the concrete. This problem applies more particularly to horizontal joints so that whereas primer is normally applied by brush to vertical joints spraying equipment of the type shown in Fig. 12. 3. has been developed for the treatment of horizontal joints. This equipment is essentially a petrol driven compressor with a spray gun mounted on a pivoted bracket so that the width of the coated surface can be varied by raising or lowering the spray gun. As additional refinements the equipment is provided with a second nozzle to provide a jet of compressed air immediately in front of the primer nozzle and for ease of cleaning a separate canister containing a suitable solvent may be connected to the spray gun after use.

The guiding principles which should be applied to the priming operation are as follows :—

(a) Do not apply excess primer. (On one occasion the author discovered that primer was applied using a conventional watering can ! Not only did this cause adhesion failures but also, during hot weather, the primer exuded from the joint onto the surface of the concrete !)

(b) Ensure that the primer coats the walls of the sealing cavity to the full depth.

(c) Allow sufficient time for the primer to dry completely before the sealing compound is applied. Adhesion failures are certain if the sealing compound is applied to wet primer. An exception to this rule is the use of a polyamide tar oil/epoxy primer.

SEALING

The correct methods for the preparation and application of proprietary sealing compounds are usually issued by the manufacturers and it is important that these recommendations be strictly observed. In the absence of reliable information on the use of compounds within the range of materials described in Chapter VIII the following general recommendations will be found of value in ensuring that the compound is used in the best possible manner :—

Hot-poured compounds. Careful control is necessary in the preparation and application of all hot-applied sealing compounds but although damage may occur when a compound is in a molten condition, the most hazardous period is the actual melting process. During this period a large amount of heat must be transferred to the compound,

M

which, being in a solid state, cannot readily be stirred to prevent local overheating. A number of precautionary measures may be taken to minimize the danger of initial overheating as follows :—

    (a) Using a wetted axe or heated spade cut the compound into pieces smaller than 6 inch cubes.

    (b) Use a heating system which can be readily controlled i.e., gas or oil burners rather than solid fuel.

    (c) Heat the compound as slowly as is practicable until there is sufficient volume of molten material to transfer heat to unmolten material by convection.

    (d) Do not at any time apply fierce heat direct to the vessel containing the compound.

    (e) At the end of a working period empty the melting vessel. If a large volume of compound is allowed to solidify in the melting vessel, local overheating is almost certain to occur when re-melting and in addition, the melting process will take a longer time.

FIG. 12. 4. An oil-jacketed melting and pouring unit for hot-poured sealing compounds.

Photograph by permission of Expandite Limited, London.

The most satisfactory equipment for melting hot-poured sealing compounds is a gas-or oil-fired melter in which a heat transfer medium is interposed between the burner and the vessel containing the compound. The most efficient heat-transfer medium is oil and several types of melter have been developed in the U.K. and elsewhere which incorporate an oil jacket around the melting vessel and may, in addition, have oil-bearing tubes passing through the vessel. These melters should be efficiently lagged to minimize heat loss. With suitable lagging it will be found that whereas it may take 6 - 8 hours to reach the optimum working temperature in a large capacity melter, it may be allowed to stand overnight and providing that the equipment is in a sheltered position, the compound will still be molten in the morning. Only a relatively small heating period will then be required to reach the desired pouring temperature.

It has been found[50] that mechanized stirring materially assists in preventing the molten compound from being overheated and also reduces the time necessary to reach the optimum pouring temperature.

Having heated the compound to the appropriate pouring consistency it should be transferred to the joint with minimum loss of temperature. If the compound is allowed to cool before reaching the joint it may not develop satisfactory adhesion. Pouring cans, if used, should therefore be lagged and the joints should be filled as expeditiously as possible with minimum delays between emptying and re-filling the cans. Alternatively small pouring machines such as that shown in Fig. 12. 4. may be used if space and accessibility permit. These machines are oil-or gas-fired and heat transfer is effected by an oil jacket and oil bearing tubes. Further refinements which can be obtained are forced oil circulation, thermostatic control and pressure ejection of the compound by means of a submersible pump.

Hot-applied strips. Preparatory heating is applied directly to strip form materials by means of a blow-lamp or gas torch and great care is necessary to melt the surfaces of the strip without causing charring. While the surfaces are in a hot molten condition the strip should be pressed firmly into the sealing cavity so that the surfaces make good contact with the concrete as illustrated in Fig. 10. 2 (b) and (d). This is a relatively simple operation in triangular cavities but although an oversize strip may be used in rectangular cavities to promote good contact, the degree of contact at the base of the cavity cannot be ascertained. Rectangular sealing cavities should not, therefore, be sealed with strip form materials which are applied hot.

Having set the strip in the joint, the compound should be heated by hot irons which will fit into the sealing cavity and can be used to press the compound against the walls and base and finally to form a

FIG. 12. 5. A hand pressure gun for applying building mastics.

smooth surface finish. In water-retaining structures in particular the compound should also be used to fill any large irregularities in the surface of the concrete adjacent to the joint in order to minimize the danger of water passing through the concrete and by-passing the seal.

Heated pastes. The most reliable method of heating sealing compounds supplied in a paste consistency is by standing the full container in boiling water, a period of one hour usually being sufficient to render the compound workable. These compounds must be applied with great care, preferably in layers, in order to ensure that air is not trapped within the sealing cavity. The application of hot irons, as used with strip form compounds, may be desirable under some circumstances. Trials have been carried out to assess the value of mastic guns for applying these materials and this method of application warrants further study and development because in increasing the rate of application it enables the compound to be applied at its optimum consistency.

Building mastics. These materials can be applied by trowel or by means of a gun which allows the material to be extruded under pressure through a nozzle as shown in Fig. 12. 5. For gun application the sealing compound is packed in cellophane, cardboard or polythene cartridges and the correct procedure to be adopted when using a gun in which pressure is exerted by hand through a pistol grip lever mechanism is as follows:—

(a) Remove the cap and nozzle assembly from the gun.

(b) Place the cartridge in the barrel and remove the cap in the case of a cardboard cartridge, or cut off the end of the cellophane cartridge and open it out to expose the mastic.

(c) Replace the cap and operate the pistol grip until it offers resistance.

(d) Holding the gun barrel at an angle of about 70° to the surface of the concrete squeeze the pistol grip and extrude the compound into the joint. As it fills the joint the rate of movement of the gun and the pressure exerted on the pistol grip should be adjusted so that a small " wave " of mastic precedes the nozzle as it is drawn along the joint. The gun should not be drawn along the joint too rapidly and to obtain a clean finish the pistol grip should be operated smoothly, reducing the rate of movement at the end of travel of the trigger and resuming the original rate after the trigger has been released and re-engaged.

(e) When a joint ends in a surface at right angles to it, the application of mastic should be stopped short and the remainder of the joint should be filled starting at the adjacent surface and working towards the compound already applied.

When a considerable length of joint is to be treated and other conditions permit, guns of the types shown in 12. 6 (a) and (b) which are operated by air pressure are justified. These air-pressure guns may be run from a small compressor, capable of delivering for example $1\frac{1}{2}$ cubic ft. of air per minute at a pressure of 30 to 80 lb./in.^{2} These guns permit very rapid application and greater precision in filling the joint.

181

Building mastics should not be applied at temperatures below 40°F. and the sealing cavity should not be wider than ½ inch nor deeper than ⅝ inch.

Emulsions. The application of emulsion type materials is relatively straightforward and may be carried out in damp conditions. Depending on consistency these materials may be poured or trowelled into the joint, the usual precautions being taken to ensure that air is not trapped within the materials during application. In general,

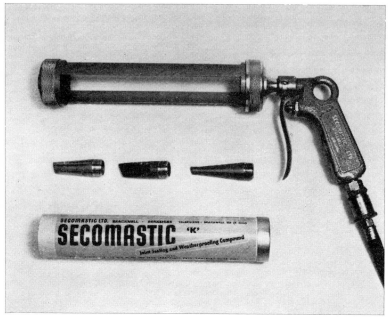

FIG. 12. 6.(a)   Air pressure gun for the application of building mastics.
Photograph by permission of Secomastic Limited, London.

however, emulsions are very sensitive to atmospheric conditions. They must never be exposed to frost otherwise coagulation will occur and when incorporating the setting agent it may be necessary to gauge the amount of the agent according to the humidity and temperature conditions. It is desirable, therefore, that the operators using these materials be guided initially by experienced technicians and be sufficiently responsible to act on their own initiative in proportioning and mixing the materials.

Adhesive preformed strips. These compounds may usually be applied by simply pressing them against a surface which is to form one of the walls of the joint and then pressing the second surface

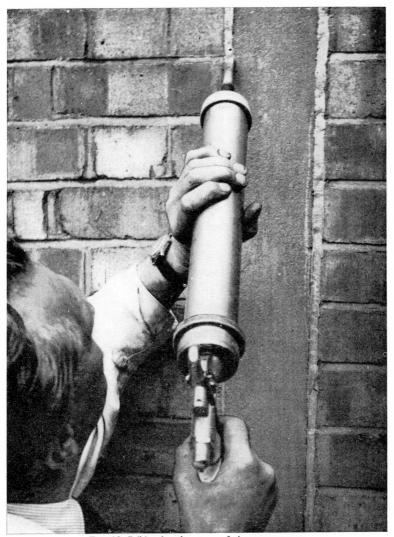

FIG. 12. 7.(b) Another type of air pressure gun.

up to it. When resinous compositions are used the two surfaces of concrete must be dry and free from dust and in the case of rubber/bitumen compositions the application of a solvent-based primer or bituminous building mastic to one or both faces of the concrete is desirable immediately before the joint is made in order to obtain optimum adhesion.

Setting synthetic resins. In the light of the limited experience on the preparation and application of these compounds it would appear that the use of infallible automatic equipment should be used. This equipment should apportion, mix and extrude the compound and thereafter the method of application is identical with that employed with building mastics except that the need for a dry surface may not be important because these compounds may displace or mix with water.

In the event of chemically setting materials being supplied in a refrigerated state it is important that they be raised to ambient temperature before being applied to the joint otherwise condensation will occur at the walls of the joint and this will lead to adhesion failures.

GENERAL CONSIDERATIONS. Summarizing the general requirements of good jointing techniques the following points may be stressed :
  (a) The walls of the joint should be free from loose dust and laitance.
  (b) The surface of the concrete should be dry
  (c) Primer, if used, should be applied as a thin film and should be allowed to dry thoroughly except in the case of cold-applied bituminous strips and polyamide/epoxy primers.
  (d) The sealing cavity must be completely filled with sealing compound.
  (e) Hot-applied compounds should be applied at the optimum working temperature and should not be overheated during preparation.
  (f) Two or three part materials which are mixed on site must be accurately apportioned and thoroughly mixed.

If all these precautions are observed the problem of sealing joints is resolved to the choice of a suitable sealing compound and whereas this choice is governed by various practical considerations, soft rather than hard compounds should be chosen whenever possible in the interests of good adhesion and ease of application.

CHAPTER THIRTEEN

Common Defects

Probably the most common defect in concrete structures is the development of cracks at construction joints. In addition to this and other defects which are related to joints and will be described in detail in this Chapter, however, random cracking, disfigurement and structural failure may be due to effects which cannot be overcome by the provision of joints. The most obvious type of failure which in some instances could be averted by the provision of joints is that due to loading. If a structure is too light to accommodate service loads, however, or if insufficient attention has been paid to the foundations, the treatment of cracks, if such treatment is possible, will be a constant care until the structure attains equilibrium. Whereas, therefore, under certain circumstances, it may not be possible for an accurate assessment to be made of the strength requirements of a structure, it is false economy to take risks deliberately in the initial design and construction when subsequent failures may be difficult or impossible to overcome. Other typical defects which cannot be corrected by the provision of joints are as follows :—

(a) Surface shrinkage cracks. This type of cracking is due to the concrete at the surface being too wet when finished or due to inadequate curing. Cracking due to either of these causes takes the form of numerous parallel disconnected cracks of the type shown in Fig. 1. 1. In some types of structure these cracks may be sealed by the application of a thin bituminous membrane but in pavings and other areas carrying traffic, full protection can be obtained only by a thick waterproof layer or possibly an expensive synthetic (epoxy) resin treatment.

(b) Cracking over reinforcement. The development of fine cracks at a close spacing is normal in reinforced concrete members but if cracking due to tensile stress concentrations is suitably controlled by increasing the steel/concrete ratio or

FIG. 13. 1.   Shrinkage cracks in concrete.   Note porosity and efflorescence due to water seepage (*top right*).

Fig. 13. 2.   Fracture and displacement of a parapet wall due to the expansion of a concrete roof.

providing contraction joints, these cracks present no hazard. Serious cracks and spalling of the surface, however, will occur if water reaches the reinforcement due to inadequate concrete cover or to porous concrete. In addition to the controls which should be applied to the fixing of reinforcement and the manufacture and placing of the concrete it is essential that the specified cover for the reinforcement be adequate to allow sufficient tolerance in fixing which cannot, by any means, be considered an operation in which a high degree of accuracy can be obtained.

(c) Frost damage. Expansion effects due to water freezing within the concrete before it has set will cause spalling and cracking. If a water/cement ratio higher than 0.5 has been

FIG. 13. 3.   Cracking due to unequal settlement.

used similar damage may occur in hardened concrete.   In order to minimize the risk of damage due to frost a water/cement ratio lower than 0.5 should be used and concreting work should preferably be stopped when the air temperature reaches 38°F. on a falling thermometer.   The use of small proportions of an accelerator together with protective coverings and heated ingredients will, however, permit concreting to be carried out successfully at much lower temperatures than this.   Moreover by incorporating an additive which entrains small air bubbles in the concrete during mixing, the expansion stresses in hardened concrete due to the effects of frost can be reduced sufficiently to prevent damage.[51]

(d)   Sulphate attack.   This type of defect is greatest in industrial areas and is caused by the formation of insoluble salts at the

188

FIG. 13. 4.   Corner cracking of a paving slab and excessive flow of a sealing compound.

surface of the concrete through combination with sulphur dioxide from the atmosphere.   The effect of this reaction is expansion of the surface layer which leads to cracking and disintegration.

Defects which are directly related to the design, construction and treatment of joints may be considered under these headings as follows :—

DEFECTS DUE TO DESIGN

There are five common types of defect which are a direct result of inadequate design.

    (a) Inadequate joint spacing.   It has already been suggested that providing appropriate steps are taken to control cracking, e.g. the installation of contraction joints and the use of heavier reinforcement in tensile zones, any cracks which do develop are of little consequence.   Failure to adopt such measures increases the possibility of cracks developing at construction joints as shown in Fig. 13. 1.

It is equally important, however, that a true assessment be made of the expansion of one part of a structure in relation to another part. Failure to provide space for expansion of flat concrete roofs is a common oversight and results in deflection and fracture of the end walls or displacement of the parapets as shown in Fig. 13. 2. This defect can be prevented either by providing expansion joints at regular spacings in the roof or by ensuring that the roof is free to slide over the end walls, in which case it carries the parapets.

(b) Differential settlement due to unequal loading conditions or load-bearing properties of the subgrade. If no allowance is made for this movement cracks of the type shown in Fig. 13. 3 may develop. In some types of structure this type of defect will have serious results because it may be progressive.

(c) Inadequate strength. A typical example of the strength requirements of a concrete structure not being satisfied either by the concrete or by the foundation is corner cracking of paving slabs as shown in Fig. 13. 4. This defect is also serious because it can be remedied only by complete reconstruction or the application of a further layer of concrete or bituminous surfacing material. Additional reinforcement in this area is a solution.

(d) Incorrect location or size of joints. Sympathetic cracking due to staggered joints is an example of the incorrect location of joints. The development of cracks at reduced sections, e.g. from the perimeter of a gulley or manhole in a road to a nearby expansion or contraction joint can often be observed and could be avoided by installing the joint in line with the break in the paving.

(e) Defects due to the failure of jointing materials such as loss of adhesion of a sealing compound or exudation of impregnant from a joint filler may be due to the size of the joints not being adequate in relation to the movement occurring and the specified performance requirements of the materials.

(f) Inadequate load transfer. In pavings, inadequate, or the omission of effective load transfer results in differential settlement and poor riding quality. In structures a tongue and groove must be substantial and should not be deeper than its width in order to avoid fracture under load. Differential movement due to lack of load transfer may lead to premature failure of

190

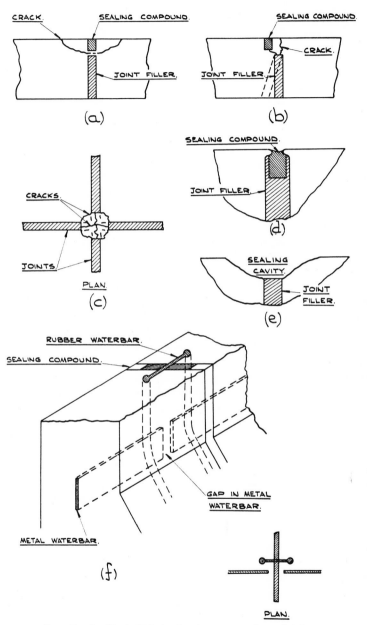

Fig. 13. 5. Typical faults in the construction of joints.

the sealing compound particularly under the high rates of strain caused by traffic loading on pavings.

DEFECTS DUE TO CONSTRUCTION METHODS

Incorrect construction procedure may be the direct cause of defects in a structure or it may result in defects due to the failure of jointing materials. The most common defects which originate from haphazard construction methods will be considered roughly according to the frequency with which they have been encountered by the author.

(1) *Cracked construction joints.* Cracks of the types shown in Fig. 13. 1. will develop at badly prepared construction joints even though suitable precautions may have been taken to accommodate high stress concentrations in the design of a structure. The preparation and formation of construction joints should therefore be very closely controlled. In difficult locations provision should be made for sealing these joints either by the use of a surface sealing compound or a water-bar so that if cracks develop they do not impair the appearance or the performance of the structure.

(2) *Discontinuous joints.* Three examples of faulty joint construction in which concrete spans the joint are shown in Fig. 13. 5. (a), (b) and (c). These faults are most common in pavings but they can also occur in other types of structure. They are all serious because they provide easy entry for water and there is no completely reliable or satisfactory remedial treatment. Incorrect location of the sealing cavity as shown in (b) can be due to displacement or distortion of the joint filler or simply to inaccurate marking off. The effects shown in (a) and (c) can be remedied when the joints are being prepared for sealing if cracks have not already developed.

(3) *Badly formed sealing cavities.* Badly formed sealing cavities may induce premature failures in the sealing compound. The formation of lips at the top of the cavity as shown in (d) makes cleaning an almost impossible task and usually a layer of joint filler or forming strip is left adhering to the walls thus preventing contact between the sealing compound and the concrete.

A spalled sealing cavity as shown in (e) is usually due to careless or premature removal of the formwork. This effect is most common in vertical joints and reduces the possibility of good contact and thus adequate bond strength between the sealing compound and the concrete.

(4) *Poor compaction.* The production of dense well-compacted concrete is one of the most important aims in nearly every type of structure. It is an essential requirement in the vicinity of all joints which must be sealed against the penetration of water. The mix proportions of the concrete must be chosen in relation to the methods

FIG. 13. 6.   Honeycombed concrete underneath a waterbar.

of compaction and the intricacy of the formwork in order to facilitate the production of dense concrete in these areas.   The installation of a waterbar, whilst making some allowance for permeability, relies on the concrete being dense and waterproof a few inches from the face of the joint.   Particular care should be taken at horizontal joints to avoid deep " honeycombing " underneath the waterbar as shown in Fig. 13. 6.   Extensive porosity will undoubtedly result in leaks and if, as in the structure in the illustration, external water pressure exists, remedial treatment using surface sealing compounds can be carried out only when suitable drainage or pumping systems have been installed to allow the concrete to dry out.

(5) *Misalignment of dowel bars.*   Two types of defect may occur due to dowel bars not being accurately aligned across the joints in a

193

N

concrete paving. When subgrade restraint is high, cracks similar to those shown in Fig. 13. 7. may develop 12 to 18 inches from the joint. If the subgrade restraint is low the movements due to moisture and temperature variations of the concrete may be concentrated at adjacent joints and extrusion or premature failure of the sealing compound may occur at these joints due to excessive variations in joint width.

FIG. 13. 7.   Subgrade failure due to a leaking joint.

DEFECTS DUE TO JOINTING METHODS OR MATERIALS

By far the most common defect in joints is loss of adhesion between the sealing compound and the concrete. In water-retaining structures and pavings this defect usually results not only in water passing through the joints but also structural failure of the type shown in Fig. 13. 7. due to its deleterious effect on the bearing properties of the subgrade.

(1) *Loss of adhesion of the sealing compound.* This defect is most likely to occur due to one or a combination of the following factors :—

    (a) Excessive vertical or horizontal movement between adjacent sections of concrete.

    (b) Application of the sealing compound to damp or contaminated sealing cavities, e.g. coated with laitance or a thin skin of fibreboard.

194

(c) The application of too much or unsuitable primer.

(d) Insufficient time being allowed for the primer to dry.

(e) Application of a hot-applied sealing compound below its optimum working temperature.

(2) *Fracture of the sealing compound.* Brittle fracture of a sealing compound is a function of rate of strain and temperature. It can occur in a given compound due to excessively high rates of strain at a moderate temperature or due to low rates of strain at very low temperatures. In both cases it indicates that the sealing compound in the joint is too hard or has a high temperature susceptibility. These shortcomings in the sealing compound may be due to one or more of the following causes :—

(a) Unsatisfactory composition.

(b) Degradation of a hot-applied compound during heating or application.

(c) Excessive loss of solvent from a solvent-based cold-applied compound due to high ambient temperatures or to very porous adjacent surfaces.

(d) Incorrect proportioning or mixing of a two-part cold-applied compound.

(3) *Flow of the sealing compound.* This defect is usually important only in its effect on the appearance of a structure but excessive flow, particularly in vertical joints, may result eventually in water penetration. Excessive flow of sealing compound in horizontal joints such as that illustrated in Fig. 13. 4. is usually due to the use of a material which is too soft or rendered soft by degradation during application. It may be aggravated by extrusion or overfilling the sealing cavity.

In vertical joints, excessive flow or slump is due to the use of unsatisfactory materials or sealing cavities which are too wide. It is suggested that vertical sealing cavities be formed not wider than $1\frac{1}{2}$ inches for hot applied bitumen/asbestos or rubber/bitumen compounds or $\frac{1}{2}$ inch for cold applied building mastics.

(4) *Cavitation.* Premature adhesion failure may be induced in a sealing compound due to cavitation under water pressure in water-retaining structures. Cavitation will occur if the joint filler does not provide adequate support for the sealing compound or if the compound is not applied to fill the sealing cavity completely. The joint filler should therefore be thoroughly tamped into the joint before the sealing compound is applied, additional fibrous material, such as hemp or slagwool, being installed to restore the desired depth of sealing cavity. Alternatively an expanding joint filler should be used such as dehydrated cork.

(5) *Bubbling.* Small bubbles of air are almost invariably trapped

in the pores of the concrete when a sealing cavity is filled with a hot-applied sealing compound. Under the varying temperature conditions in service these bubbles may tend to grow, due to expansion of the air when the sealing compound is warm and soft during the day and extraction of more air from the concrete to maintain the inflated size of the bubble as the compound cools and hardens at night. A similar effect occurs in bituminous felt roof treatments and has been described elsewhere[52] but the suggested remedial treatment of avoiding the development of adhesion is hardly appropriate here! In jointing compounds the phenomenon continues until bubbles rise to the surface where they mar the appearance of the joints although they do not affect the sealing properties. It appears likely that these bubbles form more readily when the sealing compound is applied to damp joints or is very soft.

(6) *Spalling.* The edges of adjacent slabs in pavings will spall if grit or large stones become trapped in the sealing cavities of the joints. The effect is most pronounced in expansion joints but may also occur in contraction joints to a limited extent. All sealing compounds are likely to be penetrated by grit under the action of traffic but only those compounds which will support grit under its own weight should be used in pavings. In severe conditions a harder material should be set into the sealing compound during application to prevent damage and under normal service conditions grit-filled sealing compounds should be removed and replaced during routine maintenance.

(7) *Exudation of joint filler impregnant.* Rot-proofing impregnating materials may be displaced from a joint filler if by excessive tamping or expansion of adjacent slabs it becomes completely dense. Water continually passing through the joint at high ambient temperatures such as experienced in the tropics may also displace impregnant from a joint filler. Impregnating materials will often attack sealing compounds and it is important therefore that the main causes of exudation should be avoided by adequate joint design and spacing and careful and expeditious joint treatment.

The foregoing are some of the more common defects which occur in various types of structure. All of these defects however can be avoided by the adoption of good design, construction and jointing techniques thus ensuring the production of sound structures which retain a good appearance with minimum maintenance.

CHAPTER FOURTEEN

Inspection and Maintenance

The old adage " prevention is better than cure " is particularly appropriate in the design and construction of joints to relieve tensile stresses in concrete. Not only does this apply to joint layout, design and the choice of materials, but also to methods of joint construction and to the application of jointing materials. The importance of good supervision during the construction and treatment of joints cannot be stressed too strongly. Remedial and maintenance processes are always costly and usually much less effective than the correct treatment applied during the construction period.

Before proceeding to maintenance considerations the following summary of some of the most important aspects requiring close supervision during the initial construction period may be of value :—

(1) *Construction joints.*

   (a) Cleanliness i.e. all wood shavings and other foreign matter must be removed.
   (b) Thorough application of the mortar bed without the use of excess water.
   (c) Thorough compaction of the concrete in the vicinity of the joint.
   (d) Location of the waterbar if used.
   (e) Location and copious oiling of the fillet for forming the sealing cavity if this is to be provided.

(2) *Contraction and sliding joints.*

   (a) Accuracy and smoothness of the endform joinery in vertical joints.
   (b) Accuracy in locating the fillet for forming the sealing cavity in dummy joints.
   (c) Continuity of the joints. Gaps between the sealing cavity former and the formwork should be packed with wood or fibreboard to prevent " bridging."

(d) Uniform and adequate application of the sliding membrane if provided.

(e) Careful compaction of the second layer of concrete laid to form a sliding joint.

(f) A waterbar having increased flexibility such as a hollow bulb at its centre must be located such that the joint in the concrete cuts the flexible area of the waterbar.

(g) The period of time allowed to elapse between placing the concrete and the start of joint sawing operations must be carefully assessed. This is the most difficult problem in this method forming joints and there is a definite need for a suitable testing method, such as ultra-sonic testing, which can be applied to concrete in situ.

Usually more than one saw will be required to expedite the operation.

(3) *Expansion joints.*

(a) The joint filler must be accurately cut to fit tightly against the formwork and the waterbar if present. If more than one piece of filler is used in a joint the pieces must be butted together tightly in order to prevent concrete " bridges."

(b) If a fillet is used for forming the sealing cavity it should be located firmly against the edge of the joint filler.

(c) The concrete should be placed and compacted thoroughly around the sealing cavity former.

(d) When the joint filler is placed in advance of the concreting operation as, for example, in pavings, it must be well bedded onto the formation and firmly fixed perpendicular to the formation and the side forms.

(e) In suspended joints the joint filler should be provided with a suitable mechanical fixing device such as copper nails to be set into the concrete during placing.

(f) Dowel bars, where incorporated, must be accurately located and the membrane on the sliding end should not be too thick.

(g) Waterbars, when used must be firmly and accurately anchored.

MAINTENANCE

The success or failure of maintenance work is greatly influenced by the time lag between the initial failure and the start of the maintenance operations. The development of cracks and the failure of the jointing compounds in concrete structures almost invariably lead to progressive deterioration if remedial treatment is not applied. In some types of structure such as service reservoirs and various tanks

FIG. 14. 1.   A neglected joint in a bridge deck.

and channels in sewage works, frequent or regular inspection is often impossible but many other types of structure are always readily accessible and defects can be detected and treated soon after they develop. This does not mean, however, that it is necessary or even desirable for the structures to be thoroughly examined at weekly or monthly intervals, because this would hardly be an economical approach. The tendency for defects to develop can be focussed to certain periods during the year or the age of the structure and inspection and treatment during these periods can materially affect the subsequent performance of the structure.

Many of the common defects e.g. due to shrinkage effects and faulty construction can be detected within a few weeks or even a few days of the work being carried out. It is very desirable, therefore, that all structures be closely inspected before they are put into service so that defects may be remedied with minimum inconvenience.

Subsequently, defects may be anticipated during, or immediately after, periods of extreme duress such as the following :—

(a) Prolonged exposure to very high ambient temperatures. This may cause excessive extrusion or flow of jointing compounds, displacement of sealing compounds under loading or buckling and fracture of the concrete.

(b) Prolonged exposure to very low ambient temperatures. Contraction cracks in concrete and adhesion and brittle fracture failures in joint sealing compounds may often be detected when snow is melting after a spell of cold weather. Under these conditions the presence of free water makes the failure more obvious.

(c) Autumn and Spring, when, particularly under clear sky, the daily variation in ambient temperature is very large and results in correspondingly large longitudinal and warping movements of the concrete. The vertical deflection of paving slabs due to traffic loading will also reach a maximum value during this period.

Whenever possible, therefore, inspection should be made during these periods and defects should either be treated immediately or in the event of the defects being observed during cold or damp conditions they should be marked for subsequent treatment. Neglected joints such as that shown in Fig. 14. 1. are not only costly, but also extremely difficult, to repair effectively.

Maintenance procedure. The most opportune time of the year in which to carry out maintenance work is during dry spells of weather in the Spring. At this time of the year the width of joints and cracks will be intermediate between the Summer and the Winter conditions and subsequent strains on the sealing compound and joint filler will be equally divided between extension and compression.

When cracks or failures in jointing compounds develop as a result of settlement, appropriate remedial measures such as jacking or pressure grouting should be adopted whenever possible in order to restore the equilibrium of the structure before the cracks or joints are repaired. If this is not done the failures will recur and usually progressive deterioration will follow.

If extensive damage has already occurred due to settlement or loading, partial reconstruction will often be necessary or in the case of pavings a continuous bituminous surfacing may be applied. In the latter case there is a tendency for the surfacing to crack where large strains continue to occur and it is important that these cracks be treated in the same manner as will now be described for isolated and regular cracks in concrete.

(i) *Cracks.* Isolated and well defined cracks in vertical surfaces should be chased out using a portable power-driven carborundum saw so as to provide a substantial cavity for subsequent filling with a sealing compound. In buildings and other structures not exposed to water pressure a cold-applied bitumen or resin-based mastic will usually be both satisfactory and convenient and a triangular sealing cavity $\frac{1}{2}$ inch wide at the surface and $\frac{1}{2}$ - $\frac{3}{4}$ inch deep should be formed. Conventional mastics are not generally suitable for use in water-retaining structures or in pavings, however, and in these structures a larger sealing cavity will be necessary to accommodate a hot-applied bituminous sealing compound or, alternatively, an elastomeric sealer must be used.

A machine which may be used for forming a sealing cavity in cracked pavings is shown in Fig. 14. 2. It consists of six hardened steel cutters which are free to rotate on axles spaced around the circumference of a drum driven through V-belts by a petrol engine. In the illustration the machine is being used to re-face the edges of a joint but when used for chasing out cracks a second handlebar bracket should be provided at the front of the machine to enable a second operator to guide the cutting head along the crack, the alignment of which is usually irregular.

Having formed the sealing cavity the rest of the operation is the same as that described earlier for the treatment of joints except that more care will be necessary, particularly in vertical cavities, to ensure that the sealing compound makes good contact with the concrete and also fills the cavity completely.

In certain types of structure, for example roofs and pavings, fine cracks i.e. up to 3/16th inch wide at the surface, may be sealed with a cold-applied bitumen or preferably rubber/bitumen emulsion. This sealing material can be used as a continuous waterproofing membrane on roofs and is often applied to new work in two or three coats with fibreglass reinforcement between successive layers prior to the application of an insulating or reflective top dressing. A continuous membrane is not usually suitable for the treatment of cracks in pavings and these are sealed individually, cement, limestone dust or sand being

FIG. 14. 2.   A rotary grooving machine being used to re-face the edges of a longitudinal joint in a road.
Photograph by permission of Expandite Limited, London.

spread over the emulsion immediately after application to accelerate setting and permit the passage of traffic. It is frequently found that the first application of emulsion penetrates to the bottom of the paving and if this occurs it is necessary to build up the depth of sealing compound in layers allowing a drying period of 4 to 12 hours depending on weather conditions between each successive layer.

(ii) *Joints.* The maintenance of joints is one of the very small number of regular tasks necessary to ensure the continued good performance of a well designed concrete structure. It nearly always involves the removal and replacement of those jointing materials

202

which are accessible, namely the sealing compound and the joint filler. The first stage of this operation is to remove the sealing compound and all grit which has become packed in the joint. In pavings the grooving machine shown in Fig. 14. 2. can often be used for this purpose but this machine may not always be satisfactory because it inevitably increases the width of the sealing cavity and also during hot weather it may become jammed by the sealing compound lifted from the joint. Manual or mechanized raking is usually the simplest and most economical method of removing the old sealing compound. Many plough type grooving machines constructed as trailers or as self-propelled units have been developed for use on pavings but little or no progress has been made to simplify the treatment of vertical joints which obviously present difficult problems.

Having removed the greater part of the old sealing compound and foreign matter from the joints, the walls of the sealing cavity should be cleaned as thoroughly as possible by wire brushing or blasting. If a different type of sealing compound is to be used, for example a fuel resistant compound in place of a bitumen based compound, all traces of the old material must be removed. This presents difficulties but providing that the preliminary cleaning operations are carried out conscientiously good results can usually be obtained by washing with a solvent containing 5 to 10% of a suitable detergent (paraffin has also been used) and then hosing down with water one or two hours afterwards, and brushing with stiff brushes. The joints must then be allowed to dry out thoroughly.

The joints should then be packed with suitable fibrous material such as slag wool or hemp or with cork granules to give a sealing cavity of the correct depth for the type of joint or sealing compound to be used. The application of the sealing compound and any necessary primer should then be carried out in the usual way.

Attempts have been made to repair joints in which the sealing compound has failed to adhere, or has fractured, by applying bituminous emulsions or re-heating the compound in situ but neither of these treatments is likely to provide a permanent solution.

(iii) *Water pressure within the joint.* When a sealing compound is displaced from the sealing cavity by water pressure within the joint a satisfactory repair can be effected by the use of a quick-setting cement mortar which will, however, restrain subsequent expansion. Alternatively if the water can be dispersed long enough for the concrete to dry out a conventional sealing compound may then be applied and it should be supported by a metal plate fixed to the concrete on one side and spanning the sealing cavity at the surface. The former treatment requires a considerable amount of skill and experience and is best carried out by contractors or operators having specialized knowledge of this type of work. Both treatments are costly and stress the need for careful consideration being given to the incorporation of a suitable type of waterbar when the structure is being built.

(iv) *Spalled construction joints.* The treatment of spalled joints in concrete floors is a common maintenance problem in factories, particularly when goods are transported on trolleys having hard and small diameter bogey wheels, as is often the case. Under these conditions spalling inevitably occurs where cracks develop due to shrinkage of the wearing course of concrete and progressive deterioration takes place. Conventional hot-applied sealing compounds are not entirely appropriate for the treatment of this type of defect because being fluids they do not provide adequate protection for the concrete. The two most promising treatments at the present time are the following :—

(i) Preformed rubber/bitumen strip. This material may be applied to the cracks in a triangular section to fit into the cavities formed by the spalling or it may be bonded to the concrete in the form of a thin layer which is wide enough to span the cracks. The cracks should be thoroughly cleaned before the strip is applied and an adhesive bond may be achieved by softening the strip with a suitable solvent such as white spirit or by the application of heat immediately before application.

(ii) Synthetic elastomers. These materials have not been used extensively for this type of work but as they do not require heating and can also be applied by gun they should be particularly suitable. Moreover they can be formulated to set to an elastic consistency which will protect the concrete. In addition to the usual cleaning procedure the concrete should be degreased before these materials are applied.

The foregoing maintenance methods have been described in relation to the most common defects which may be encountered but the general principles apply to all maintenance work. The essential requirements are thorough preparation, the use of a sufficient quantity of the correct material and the adoption, whenever possible, of measures to eliminate the cause of failure.

CHAPTER FIFTEEN

Testing Methods

It is not proposed in this chapter to describe in detail any particular series of tests which may be applied to jointing materials but merely to refer to a selection of typical test methods which are in regular use. The source of any method which has been adopted as a National standard will be given to enable the reader to obtain complete details of the testing procedure if required.

THE JOINT FILLER

The most comprehensive range of testing methods for cellular joint fillers are almost certainly the U.S.A. Federal Specification HH-F-341A[53] and A.S.T.M. D1751 and 1752. These specifications have been adopted with minor modifications and additions by a number of Authorities outside the United States and in the light of present knowledge it seems likely that they will continue to be an effective control for some time. The following types of joint filler are covered by the specifications:—

Cork	
Sponge rubber	Type I Class A—highly resilient
Cork Rubber	
Bituminous fibre	Type I Class B — moderately resilient.
Glass fibre	
Self-expanding cork	Type II — self-expanding.

The characteristics of these joint fillers are assessed according to the results of performance tests carried out in accordance with U.S.A. Federal Specification SS-R-406[54], appropriate limits being set for each type and class of material as follows :—

(i) *Compression and recovery*. The pressure required to compress an accurately cut specimen $4\frac{1}{2}''$ x $4\frac{1}{2}''$ to 50 per cent. of its original thickness without the application of lateral restraint shall be not more than 1,500 lb./in.^{2} After two repetitions of this loading at a rate of approximately 0.05 inch per minute, allowing a 30 minute recovery period between each repetition, the recovery of the specimen one hour after the final loading shall be not less than 90 per cent. in the case of Type I Class A and Type II materials and not less than 70 per cent. for Type I Class B materials. Self-expanding cork shall be subjected to the Expansion Test and then allowed to dry for 24 hours before being submitted to this test and the Extrusion Test.

(ii) *Loss of impregnant.* Bituminous fibre fillers shall be mounted between waxed paper in the Compression Test and after completion of the test the loss in weight of the specimen shall not be more than 3 per cent. of its original weight.

(iii) *Extrusion.* A specimen identical with that used for the Compression and Recovery Test shall be compressed to 50 per cent. of its original thickness confining lateral movement to one edge. The amount of extrusion of the free edge shall not exceed 0.25 inch.

(iv) *Expansion.* Self-expanding cork specimens for use in the Compression and Recovery Test and the Extrusion Test shall be immersed in boiling water for one hour and after cooling at room temperature for 15 minutes they shall have a thickness not less than 140 per cent. of their original thickness.

(v) *Insolubility.* Type I Class A and Type II materials shall show no signs of disintegration after immersion in boiling hydrochloric acid for one hour.

(vi) *Weathering.* This is an optional additional test in which two specimens of joint filler are exposed to a temperature of 74°C. (165°F.) for a period of 7 days and then subjected to ten cycles of freezing and thawing in a saturated condition prior to the application of the Compression and Recovery and the Extrusion Tests.

(vii) *Site performance.* A clause is included in the Specification to cover, in general terms, the robustness of the joint filler so as to ensure that it will withstand twisting, bending and exposure to atmospheric conditions on site.

THE SEALING COMPOUND.

In view of the importance of the functions to be fulfilled by the sealing compound it is very desirable to ensure that the properties of the compound are adequate to fulfill these functions. As with joint fillers, however, it would be wrong to specify a sealing compound by controlling its composition because such a step would inhibit the development of new and possibly more suitable materials, which may also be more economical in initial cost. Once again, therefore, most specifications control the quality of compounds on the basis of performance tests which bear some relation to conditions met in service. This method of specifying materials also has the advantage of being open to wide modifications and in the case of joint sealing compounds, frequent modifications to specifications are necessary to meet the changing needs of new design and construction methods. The performance requirements for sealing compounds have become progressively more exacting in the past and will probably continue to do so in the future.

Hot-poured compounds. The main clauses in a few typical Specifications and testing methods which are applied to hot-poured joint sealing compounds are as follows :—

(a) U.S.A. Federal Specification SS-S-164.[55]

   (i) Pour point. The compound must be readily pourable at a temperature at least 20°C. below the temperature at which deterioration can be detected by the flow test.

   (ii) Consistence. The penetration of a cone, the dimensions of which are given in SS-R-406[54] and in the Institute of Petroleum Testing method IP50/56[56], loaded to 150 grams must not exceed 0.90 centimetres in 5 seconds at a temperature of 25°C. (77°F.).

   (iii) Flow. A plaque of sealing compound 4 cm. x 6 cm. x 0.32 cm. thick is cast on a bright tin panel and after cooling and trimming to size is mounted at an angle of 75° with the horizontal in an oven maintained at 60°C. (140°F.) for 5 hours. The change in length of the specimen at the end of this period must not exceed 0.5 cm.

   (iv) Extensibility. A specimen of sealing compound is cast between 2 in. x 2 in. faces of two blocks made from coarse sand/cement mortar set one inch apart. Having been trimmed and then maintained at a temperature of −16.7°C. (0°F.) for a period of not less than four hours, the specimen is extended by an amount of 0.50 inch in an extension testing machine driven mechanically at a uniform rate of 0.125 inch/hour. After the specimen has been extended the blocks are removed from the machine and stood on end so as to return to their original positions or after two hours standing are forcibly brought to these positions at a rate of 0.10 inch/minute. The specimen is subjected to five cycles and must not, at any time during the test procedure, develop a crack, separation, or any other opening deeper than ¼ in.

(b) U.S.A. Federal Specification SS-S-167.[57]

   (i) Pour point. As described in SS-S-164.

   (ii) Consistence. In addition to the penetration test limits set

FIG 15. 1.   A low temperature extension testing machine for concrete joint sealing compounds.

FIG. 15. 2.   A typical " pass " result in an immersed extension test.

in SS-S-164 further limits are applied to assess resistance to petroleum solvents. After 48 hours immersion in a mixture of aromatic hydrocarbons at 41°C. (100°F.) followed by a drying period of one hour, the penetration of the specimen at 25°C. (77°F.) must not exceed 1.00 cm. or increase by more than 0.25 cm. from that of the non-immersed specimen.

(iii) Solubility. A penetration test specimen after immersion and drying must not vary in weight by more than 2 per cent.

(iv) Flow. The permitted increase in length of the specimen subjected to the flow test is 2 cm.

(v) Extension. The extension and compression test comprises three cycles carried out at a temperature of −10°C. (15°F). An identical test is made on a specimen immersed in the test solvent and after the third extension the total area of bare concrete exposed on the face of either block must not exceed $\frac{1}{2}$ inch by $\frac{1}{2}$ inch.

Difficulty is sometimes experienced in the interpretation of the immersed extension test due to the effect of the solvent on the cohesive properties of the sealing compound. A typical result is shown in Figs. 15. 1. and 15. 2.

(c) BSS2499.[58]

(i) Pour point. The pour point is estimated from an empirical relationship based on observations of the pouring characteristics of available compounds. This relationship is :—

$$P = 66 + S + 3.4I. \quad (°\text{Centigrade}).$$

where P = the pour point (°C.).

S = the softening point (ring and ball °C.).

I = the penetration index [59].

(ii) Consistence. There is no consistence test in this Specification but requirements similar to those in SS-S-167 are included in a modified version of BSS2499 used to cover fuel-resistant compounds.

(iii) Filler settlement. A specimen of sealing compound is maintained for two hours at the safe heating temperature recommended by the manufacturer in a sedimentation apparatus which permits samples to be taken from the top, centre and bottom of the specimen. On analysis the filler contents of the three samples must not vary by more than ten per cent. of the mean.

(iv) Flow. A sample of the sealing compound is poured into three steel moulds each 2 in. by $\frac{1}{2}$ in. wide by 1 in. deep

O

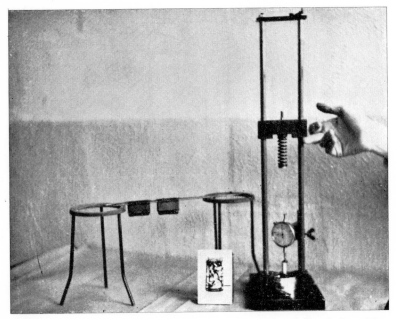

FIG. 15. 3. Performance tests for concrete joint sealing compounds. Left to right :
BSS2499 flow test, U.S.A. Federal Specification flow test, BSS2499 impact penetration test.

and having a full length slit $\frac{1}{4}$ in. wide in its base. After
cooling and trimming, the moulds are mounted on a suitable
frame and maintained at a temperature of 45°C. (113°F.)
for five hours. At the end of this period the material which
has flowed through each slit is trimmed off and by difference,
the weight of this material is expressed as a percentage of
the weight of material required to fill the mould completely.
The Specification defines three classes of sealing compound :

Class A 0 to 5% flow — Material suitable for steep
gradients in warmer districts.

Class B 5% to 10% flow — Material suitable for medium
gradients in warmer districts and may be used on
steep gradients elsewhere.

Class C 10% to 15% — Material should be used only
in joints on slight gradients.

An example of a compound having class A flow is
shown in Fig. 14. 3.

(v) Resistance to grit. A standard penetration tin is filled with
sealing compound up to $\frac{3}{8}$ in. of the top and placed in the

impact testing machine illustrated in Fig. 15. 3. A light plunger having a diameter of 1 cm. located by a guide, is rested on the surface of the compound and is given three blows with a 700 g. weight dropped vertically 40 cm. at 15 second intervals. The position of the plunger is measured 15 seconds before the first blow and the penetration after 5 minutes is recorded in 0.1 mm. as the impact penetration. Compounds are then classified as :—

> Grade A — 20 or less —. For use on roads where grit and small stones are present in very large quantities.
> Grade B — 20 to 40 —. For use on roads where grit is found only occasionally.

(vi) Extension. This test is similar to the U.S.A. Federal Specification test except that the specimen is smaller and 50 per cent. extension at the standard rate of 0.125 in/hour is applied at a temperature of 0°C. (32°F.) in the case of bitumen based compounds and 75 per cent. extension for fuel-resistant compounds.

The tests described in this Specification are also used in France and in Belgium but the following modifications have been made to the limits and test temperatures :—

France.[60]

(i) and (ii) The pour point is replaced by standard consistence tests.
(iii) Filler settlement. No limits are applied but low settlement results are recommended.
(iv) Flow. A temperature of 50°C. is used and compounds having low flow values which also pass the extension test are considered to be of particular interest.
(v) Resistance to grit. An upper limit of 10 mm. is applied without the use of grades.
(vi) Extension. An extension of 50% is applied to the American type specimen at temperatures of O°C. (32°F.) and —10°C. (15°F.).

Belgium.[61]

(i) and (ii) Pour point. No test is applied.
(iii) Filler settlement. An upper limit of 15 per cent. is applied.
(iv) Flow. An upper limit of 20 per cent. is applied without the use of classes.
(v) Resistance to grit. An upper limit of 10 mm. only is applied.
(vi) Extension. A manually operated apparatus is permitted which permits successive extensions of 0.2 mm. to be applied

every two minutes. Fuel-resistant compounds are subjected to this test whilst partly immersed in aviation fuel.

British Standard 2499 is currently in the process of being redrafted. In the new draft two types of compound are proposed in each of ordinary and fuel-resistant grades. The Flow Test has been retained with a general maximum limit of 18 per cent and the Extension Test is included using an extension rate of 0.25 in./hour and an extension of 75 per cent in two cycles at temperatures of 0°C. and −15°C. for each of the two types. A simple cone penetration test has been substituted for the impact penetration test and a degradation on heating test for the pour point and filler settlement tests. An extension test after fuel immersion has been inserted for the two fuel-resistant grades.

These modifications thus cater for the previous BS2499 type and a similar fuel-resistant grade and also for the softer type such as that covered in the U.S. Federal Specifications SS-S-164 and SS-S-167.

(d) DIN 1996.[62]

(i) Consistence. This is controlled by the standard ring and ball softening point test for asphaltic bitumen as described in the Institute of Petroleum Testing Method IP58/56.[56] The limits set are 55 to 70°C. (131 to 158°F.).

This testing method is to some extent preferable to the penetration test because it is more rapid and requiring only a small sample, it may easily be carried out on compound taken from a joint.

A further consistence test is carried out on fuel-resistant sealing compounds to assess the effect of immersion.

(ii) Flow. A specimen of sealing compound is cast in a cylindrical mould 20 mm. long and 10 mm. in diameter and it is rested on a corrugated brass sheet inclined at an angle of 15° with the horizontal. The specimen, mounted in this manner, is exposed to a temperature of 45°C. (113°F.) for 30 minutes after which period its length must not increase by more than 10 mm.

Susceptibility to deformation at high ambient temperatures is also assessed by a method based on the work of Nüssel. A sample of compound weighing 50 grams is moulded into a sphere by hand and after cooling for two hours at 0°C. (32°F.) on a sheet of glass it is placed in an oven maintained at 45°C. (113°F.) for either one hour or twenty-four hours. The ratio of the diameter of the sphere to its height after testing must lie within the following limits :—

After one hour   1.2 to 4.0

After 24 hours   1.5 to 9.0

(iii) Filler content. The filler content of the sealing compound must not exceed 50 per cent. and not more than 20 per cent. of this should be retained on an 0.2 mm. mesh sieve.

(iv) Low temperature consistence. A spherical specimen prepared in the same manner as for the high temperature deformation test is cooled to −10°C. (15°F.) and is allowed to fall from progressively increasing heights onto a steel plate 10 mm. thick. The specimen must withstand a fall of four metres without fracture.

(v) Extension. This test is similar to the American and British tests except that the specimen of sealing compound is larger (200 mm. x 100 mm. x 15 mm. thick) and the rate of extension, which is applied manually, is 1 mm./hour. The test is carried out at −10°C. (15°F.).

As a result of the increasing demand for sealing compounds having high performance characteristics the limits set in a number of tests described for hot-poured compounds are so strict as to be barely obtainable with the materials and manufacturing processes at present in use. The interpretation of test results has therefore become extremely difficult in tests which rely on visual observation only, such as the immersed extension test.

COLD LAID COMPOUNDS

The most promising line of development at the present time appears to be one-, two- and three-part cold laid sealing compounds containing polysulphide polymers (Thiokols) and other synthetic resins or bituminous materials. Bituminous materials containing Thiokol are particularly noteworthy because they are not so sensitive to the moisture conditions of concrete and the need for careful preparation of laboratory specimens, which detracts from full-scale conditions, may not arise.

Testing methods for these cold laid compounds should not differ significantly from the methods used for hot poured compounds except that a setting time must be allowed and the mixing procedure must be standardized for each compound. The adoption of the manufacturers instructions as in the American Federal Specification SS-S-170[63] is probably the best approach to the preparation of specimens in view of the varied nature of the materials involved.

BUILDING MASTICS.

At the present time there is no British Standard Specification for

FIG. 15. 4. A slump test for building mastics showing satisfactory and unsatisfactory performance.

building mastics. The preparation of a Standard to cover building mastics which are applied by gun, trowel or in strip form is, however, being studied by a Technical Committee of the British Standards Institution. Various methods of test are being examined in an attempt to set limits to physical and chemical properties of the materials on the basis of laboratory tests which ensure satisfactory behaviour during storage, handling and subsequent performance in practice.

Gun- and trowel-applied building mastics are covered by Government Specifications in the U.S.A. and Canada and whereas some of the methods of test used in these Specifications are not entirely applicable to compounds used in other countries a brief outline of the main clauses may be of some value in considering the most important characteristics requiring control.

FIG. 15. 5. An oil seepage test for building mastics showing satisfactory and unsatisfactory performance.

(a) U.S.A. Federal Specification TT-C-598[64]

   (i) Shrinkage. A sample of mastic is applied as a layer $\frac{1}{8}$ inch thick to an area $2\frac{5}{8}$ inch in diameter enclosed by a brass ring 1/32nd inch thick by $\frac{1}{2}$ inch wide on the prepared surface of a block of limestone obtained from a particular district. After determining its volume the specimen of mastic is exposed to air at room temperature for 15 days. Its volume is again determined and should not have decreased by more than 20 per cent.

   (ii) Extensibility. A layer of mastic $\frac{1}{4}$ inch thick is set between

215

a block of limestone and a sheet of glass to cover an area of $3\frac{1}{2}$ inch by 3 inch. By rotating two metal rods which separate the block and the glass, the mastic is extended by 25 per cent. and after this extension the area of contact between the mastic and the glass should be not less than 90 per cent. of the total area.

(iii) Rate of hardening. The consistence of a fillet of mastic $\frac{3}{8}$ inch wide x $1\frac{1}{4}$ inch deep x $3\frac{1}{4}$ inch long is measured using the standard penetration apparatus but with the load reduced to 12.5 grms. After 15 days exposure at 22.5°C. (70°F.) the skin is removed from the surface of the mastic and its consistence is again determined. The mastic should not harden by more than 40 per cent. and should show no signs of cracking.

(iv) Oil retention. The depth of oil penetration from the line of contact of the mastic and the limestone blocks used in the rate of hardening test should not exceed 1/16 inch when its final consistence is measured.

(v) Slump. A specimen of the mastic in a rectangular cross-section steel trough similar to those illustrated in Fig. 15. 4. is exposed vertically at a temperature of 22.5°C. (70°F.) for 24 hours and at a temperature of 50°C. (122°F.) for a further period of 24 hours. The mastic should not flow or slump more than 3/16 inch below the lower end of the trough.

(b) Canadian Government Specification No. 1-GP-23-1946.[65]

(i) Consistence. This is a general clause requiring that the mastic be uniform and homogeneous such that it may be spread easily.

(ii) Extensibility. A ribbon of mastic is applied to seal to a depth of $\frac{5}{8}$ inch the gap between two blocks of limestone held $\frac{1}{2}$ inch apart by a metal spacer. After a curing period of 40 days at 25°C. (77°F.) the volume enclosed by the ribbon of mastic and the metal spacer is filled with water containing a dye and the blocks are pulled apart at a rate of 0.3 inch per minute. The mastic should withstand an extension of 0.1 inch without permitting the passage of water.

(iii) Slumping. This test is similar to that used in the U.S.A. Federal Specification but the lower end of the trough is extended beyond the edge of the specimen of mastic and the flow of mastic along the trough is measured.

(iv) Bleeding. A sample of mastic contained in a steel cylinder

FIG. 15. 6. An example of a workability test apparatus for gun-applied mastics.

$\frac{5}{8}$ inch internal diameter and $\frac{3}{4}$ inch in height is placed on a wad of 30 Whatman No. 1 filter papers and is clamped in such a way that the mastic is in intimate contact with the paper by means of a plug fitting inside the cylinder. A similar arrangement employing a 200 gram weight to maintain close contact is shown in Fig. 15. 5. The performance requirement is that not more than 15 filter papers should be stained. When a 200 gram weight is used not more than three filter papers should be stained including the uppermost paper.

Fig. 15. 7. Results of shrinkage tests on mastics from various sources showing markedly different skin formations.

In addition to the tests described in these two Specifications, workability should be controlled by a standard extruding apparatus such as that shown in Fig. 15. 6. in the case of gun-applied mastics and a penetration test using, for example, ball-ended needles to minimize drag in the case of strip mastics and those applied by trowel.

It is also desirable that a visual assessment be made of the time required for the development of a surface skin. This assessment might also include wrinkling which, if excessive, gives an unpleasant appearance as shown in Fig. 15. 7. although it may not, in practice, have any adverse effect on performance.

ELASTOMERIC SEALING COMPOUNDS

In addition to the U.S. Federal Specification SS-S-170 for bituminous compounds containing polysulphide elastomer or Thiokol, specifications have also been drawn up for unmodified polysulphides which are usually in the same gun grade consistency as oil based building mastics. A typical example of these is American Standards Association Specification A116.1-1960. This specification provides for two grades of compound.

Class A (Self-levelling) Compounds — i.e. those suitable for horizontal joints.

Class B (Non-sag) Compounds for use in vertical joints.

The most significant requirements of this specification are extensions of 150 per cent before and after water immersion and of 100 per cent after heat ageing, cycling at −20°F. and 77°F. and also after sun lamp exposure through glass. The last of these requirements has been introduced as a result of adhesion failures occurring at the glass/ compound interface under sunlight and special primers have been developed to overcome this effect.

WATERBARS.

Waterbars are usually specified by trade names on the basis of test results published by the manufacturers or by imposing limits to existing standard methods of test. Typical requirements for copper waterbars according to British Standard Specification 1878:1952 are given in Chapter VIII.

The quality of the rubber used in the manufacture of synthetic or natural rubber waterbars is commonly controlled by the standard methods of test detailed in British Standard 903:1960[66]. Quality control is achieved by requirements of minimum tensile strength of 3,200 pounds per square inch with an associated elongation at break of 550 per cent and after ageing for 96 hours in an air oven at 100°C. a tensile strength of 2,000 pounds per square inch and an elongation not less than 65 per cent of the original.

In the case of polyvinyl chloride, more elaborate controls are necessary to establish the effectiveness and permanence of the plasticizing agents, without which the compound would become brittle. Thus reference may be made to British Standard 2571:1963, Class 3 Type G4 generally being chosen and requiring tensile strength minimum 1600 pounds per square inch with elongation at break of 225 per cent, a British Standard Softness Number in the range 45-70, cold flex temperature of −5°C. maximum before ageing and 0°C. maximum after ageing, loss of weight on heating of 3.0 per cent maximum, water absorption of 150 milligrams maximum and water soluble matter of

20 milligrams maximum. Testing is carried out to the general methods of British Standard 2782:1965 except that material used is cut from the extruded waterbar instead of using calendared sheet.

A similar and somewhat more searching series of test methods and limits can be applied by reference to American Standard Test Methods D412-62T, D1706-61, D1004-61, D1203-61T, D792-61T but in general, the properties required to satisfy the British Standard are so far in excess of any practical requirement that the application of stricter controls serves only to increase construction costs with no significant improvement in the overall performance of the structure.

One of the most important considerations in the specification of waterbar materials is the strength of welded or vulcanized joints which in turn, may be affected by the design of the waterstop profile and its uniformity. It is usual for a requirement to be inserted in the specification that the strength across a waterbar joint made by either factory or on-site methods be not less than 65 per cent of the unjoined material in the case of rubber waterbars and not less than 75 per cent in the case of polyvinyl chloride waterbars.

The foregoing descriptions of existing and projected testing methods, whilst giving only a general idea of the controls which may be applied to jointing materials, will, it is hoped, be of some assistance to Specifying Authorities and perhaps also to manufacturers. The author believes that in setting out various specifications in this manner the " comprehensive " type of Specification intended to cover a universal jointing material which does not in fact exist, may be avoided and instead a strictly controlled compromise be chosen to suit specific conditions.

CHAPTER SIXTEEN

Overseas Conditions

The construction of large concrete structures which require articulation is increasing in districts where climatic and other conditions produce often unpredictable effects. These conditions may frequently require adjustments to be made to conventional jointing methods and materials.

The greatest differences occur in under-developed territories in which there may be only limited experience of the significance of prevailing conditions.

The most important variables which require consideration are :
(1) Temperature effects.
(2) Moisture conditions.
(3) Soil conditions.
(4) Bacteria and insects.
(5) Corrosion.

Unsuitable storage facilities and the scarcity of plant and experienced labour may also present difficulties in the construction and treatment of joints using conventional European methods and materials.

Temperature Effects

The obvious and probably most important considerations affecting the design and spacing of joints and the properties required in jointing materials are the daily and annual temperature variations and the maximum and minimum temperatures. Whereas shade atmospheric temperatures are obtainable for most districts throughout the world, the surface temperatures and temperature gradients which· develop in the walls and roofs of structures are not readily available in many countries, and it is these temperature effects which are of particular interest to the civil engineer.

Valuable records of surface and air temperatures are being made by a number of research laboratories overseas, and these records will enable engineers to calculate the anticipated expansion and contraction

FIG. 16. 1. Shrinkage at an expansion joint in a roof.

FIG. 16. 2. Fracture and buckling of a vermiculite/cement screed laid over an expansion joint.

to be accommodated in structures. Such records are being prepared in the Persian Gulf, where typical maximum surface temperatures for yellow sand-lime brick walls and flat roofs where maximum shade temperatures are 120 deg. F. are as follows :

Surface :

Sand lime brick wall—East	134 deg. F.
,, ,, ,, ,, —North	125 deg. F.
,, ,, ,, ,, —West	148 deg. F.
,, ,, ,, ,, —South	127 deg. F.
Mud screed roof —	169 deg. F.

Until these records have been completed, however, the spacing and width of joints is being estimated from a knowledge of prevailing atmospheric temperatures.

ANTICIPATED MOVEMENTS

An example of the magnitude of the movement which occurs in regions around the shores of the Persian Gulf is illustrated in Fig.

222

16. 1. This joint is in the roof of a large reinforced concrete building which was completely articulated by ¾in. wide joints spaced at 60 ft. intervals.

The joints are filled with two ⅜in. thicknesses of impregnated cane fibre-board joint filler, and the contraction between adjacent 60ft. slabs, illustrated by the space between the concrete and the filler, is that due to initial shrinkage and the contraction occurring between mid-summer (shade temperature 120 deg. F.) and late spring (shade temperature 100 deg. F.).

The magnitude of the problem is even more apparent if consideration is given to the winter condition when the atmospheric temperature is in the region of 40 deg. F.

An example of the expansion occurring subsequent to winter construction in the same area is illustrated in Fig. 16. 2. This shows buckling and disintegration of a vermiculite/cement screed 3in. thick which has been applied to a roof for insulation purposes and laid over an expansion joint between two 60ft. concrete slabs.

In Japan, where the current practice is to provide expansion joints at spacings of 40-60 metres (131-197ft.), it has been found[67] that considerable variation in expansion and contractions occurs between the roof and the foundations of buildings ; the movement of foundations being 15-20 per cent. of that occurring in the roof. Measurements of daily and annual movements in fully-exposed concrete slabs 85.5 metres (280ft.) in length were 1.5 mm. (0.06in.) and 11 mm. (0.43in.) respectively.

PROPERTIES OF JOINTING MATERIALS

In addition to the modifications which must be made in the width and spacing of joints, further modifications are frequently necessary in the properties of bituminous and resinous compounds used to seal the tops of expansion and contraction joints. Light-coloured building mastics which give completely satisfactory performance in temperate climates exhibit serious flow when used in ½in. wide vertical joints and exposed to sun temperatures of the order quoted above. The remedial treatment of mixing filler with the compound on site as shown in Fig. 16. 3. is not recommended.

Bituminous materials are relatively more susceptible to these temperatures because they are black. The use of oxidised bitumen and the addition of rubber however causes these materials to be less affected than resinous materials for a given temperature range. For certain applications, some modification is required even in these materials when the annual temperature variation may be of the order of 120-140 deg. F.

A larger frequency distribution of temperatures of the same order as, or only slightly higher than, the maximum summer temperatures in temperate climates will also result in a marked difference in the performance of bituminous sealing compounds, particularly when other considerations exist. For example, it is apparent that the effects of oil spillage on conventional rubber/bitumen sealing compounds in the servicing aprons and hard standings at civil and military airports are much more severe in tropical and sub-tropical climates because the compounds are softer and are therefore more susceptible to attack over a larger period during the year.

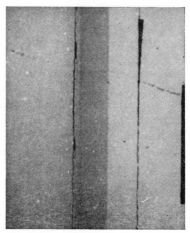

FIG. 16. 3.   The addition of filler to a building mastic on site to overcome slump.

FIG. 16. 4.   Shrinkage of precast concrete building units. Note the joint filler slipping out of the expansion joint.

MOISTURE EFFECTS

Apart from the effects of moisture variations on soils, which will be incorporated in the next section, both humidity and rainfall will influence joint design and the properties required in jointing materials.

In the desert areas of the Middle East and also in the high veldt areas of South Africa, very low humidity is experienced at certain periods of the year. These periods, which are often of long duration, present a serious problem in curing concrete.

This problem is particularly severe in desert areas where the atmospheric temperature precludes the use of certain types of curing membrane. Bituminous membranes are unsuitable because of their colour and resinous materials break down. Moreover, at the same period of the year water is frequently in short supply.

224

SHRINKAGE AND WARPING

It often happens, therefore, that in spite of elaborate precautions and continual hosing down, cast *in situ* concrete dries out too rapidly and thus excessive initial shrinkage occurs. Shrinkage and warping of precast cladding blocks is also found and as shown in Fig. 16. 4. causes unsightly cracks. These cracks can be sealed only by chasing out and pointing with a flexible sealing compound which will accommodate subsequent movement.

It has been found in the Middle East that serious cracks develop due to shrinkage effects in reinforced concrete if expansion joints only are incorporated at intervals of 60ft. When joints have been installed at spacings of 30ft. or less, however, cracking has been eliminated. Similar but probably less severe cracks have been recorded in England[68] where the contraction joint spacing was 50 ft.

A survey of cracks in reinforced concrete buildings in Japan revealed that no cracks occurred in spans of 10 metres (32.8ft.) or less. This investigation also indicated that cracks developed when a movement of 0.04 per cent. occurred due to shrinkage.

In tropical forest areas where humidity is generally high, the movements which would normally occur due to large temperature variations are in some measure balanced by movements in the opposite direction due to moisture variations.

SOIL CONDITIONS

In temperate climates allowance must often be made for movements of the subgrade either due to the effects of moisture or to mining operations. In addition, however, to the wide range of moisture sensitive soils which may be encountered in countries overseas, the prevailing climatic conditions have a very significant effect on the movements occurring in structures.

The most serious effect due to soil moisture conditions is undoubtedly expansion or heaving such as that found in Burma, Australia, South Africa, Texas, Palestine and other countries where high insolation and evaporation characteristics of the climate combine with the moisture sensitivity of certain types of clay.

Whereas in most temperate climates the expansion and contraction of soils is largely seasonal, it has been suggested[69] that irreversible expansion, under certain climatic conditions, may be a more important effect. The possible form of the two types of movement is shown in Fig. 16. 5.

Continuous heaving appears to occur in areas where rocks in varying states of decomposition are found at a considerable depth

225

below the surface and where the natural water table is also very low. When a structure is built in such an area suction effects occur beneath the foundations due to reductions in :

(a) The temperature gradient in the soil.

(b) The rate of evaporation of moisture from the surface.

Other causes for an increase in the moisture content of the soil are drainage of surface water from the building and leakage from waste pipes which have fractured as a result of the movement.

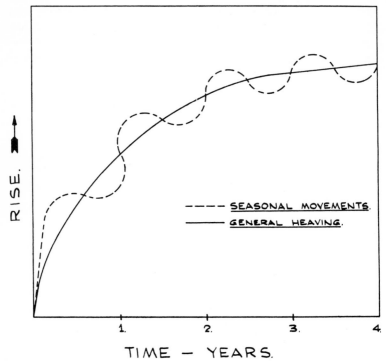

RISE.

TIME — YEARS.

Fig. 16. 5.   The form of seasonal and accumulative movement of moisture susceptible soils.

Recorded Movement

Measurements recorded[69] on a new house with drained foundations which were carried down into hard decomposed Ecca shale in the Transvaal indicated that the building had risen 2in. four years after construction.   Undoubtedly larger movements than this have occurred elsewhere and specialised methods of construction must be adopted if serious cracking of a structure is to be avoided.

Three types of construction can be employed to combat or to accommodate large soil movements as follows :

(a) *Flexible construction.* This is more particularly appropriate to buildings and is essentially curtain wall construction. The structure should be composed of a light framework carrying interlocking preformed panels which may be jointed with flexible sealing compounds to provide weather protection.

(b) *Rigid construction.* A rigid structure can be constructed either to move as a unit, rising and falling with the soil movements, or it may be constructed on foundations which extend beyond the zone in which expansion effects may be anticipated, i.e. to the natural water table.

   If the former method is adopted particular attention must be paid to the distribution of the reinforcement at reduced sections, at the bottom of walls and across construction joints in perimeter walls. Internal partition walls and floors may be provided with flexible joints to accommodate movement.

   In the second method the structure is supported on piles which isolate it completely from the anticipated soil movements. If this method of construction is adopted the grade beams and the ground floor of the structure should be completely clear of the formation.

The completely rigid form of construction is probably the most reliable, but all these three methods have been used, and it seems probable that each method may be suitable for a particular type of structure.

BACTERIA AND INSECTS

A number of the materials used for filling and sealing joints in structures contain fibres, oils or resins derived from vegetable matter, and may therefore be prone to attack by bacteria and insects. Two types of material in particular may fall into this category. These are the building mastics and preformed joint fillers.

Building Mastics

Resin-based building mastics are likely to be attacked by bacteria and insects. Some of these materials contain mineral oils and solvents which may discourage such attack, but even these may not be immune to the more robust and persistent forms of attack such as that of termites.

227

Materials of this type may also tend to slump when exposed to high atmospheric temperatures and, moreover, at these temperatures a considerable reduction in effective life may also be expected. It would seem, therefore, that the use of resin based building mastics, should, if possible, be avoided in tropical and sub-tropical climates.

Hot applied bitumen or rubber/bitumen mixtures are unlikely to be attacked by bacteria and insects, and these materials may also be formulated so as to possess the other desirable characteristics, such as resistance to flow and weathering, which cannot be achieved with resin-based materials. Furthermore, they are, in general, more economical, and this is often an important consideration in the construction of buildings in under-developed territories.

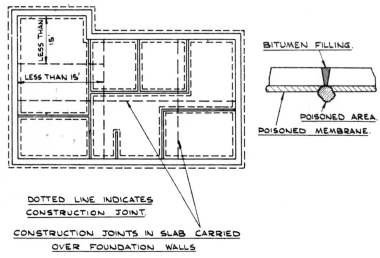

FIG. 16. 6.   Details of construction joints in concrete sub-floor.

Bitumen-based building mastics should also be suitable so far as resistance to weathering and attack by bacteria and insects are concerned, but these materials should be used only in joints up to ½in. wide to minimise flow.

Preformed Joint Fillers

Many of the cellular joint fillers comprise or contain vegetable or animal material and will therefore be prone to attack by bacteria and insects. A coating of mineral oils or bitumen will discourage this attack but may not altogether prevent damage occurring. The most effective treatment for these materials is probably thorough impregnation with pentachlorophenol, copper naphthenate or similar chemicals which are known to give protection from these agencies.

Even these materials, however, may not give complete protection from anaerobic bacteria or from marauding insects suffering from "lean times." Termites have been known to eat through the lead covering of electricity cables to gain access to the internal cotton covering which, while being comestible, would hardly appear to be sufficient incentive for such a task.

Protection of Buildings

The protection of buildings against subterranean termite attack must also be considered, and in solid concrete floors where provision for ventilation and inspection cannot be made, the concrete itself must act as a termite barrier. It has been stated[60] that cracks as fine as $\frac{1}{16}$ in. are sufficient to provide an entry for termites. The most common cause of cracking is shrinkage of the concrete on setting, and it has been suggested that joints should be provided at 15 ft. spacing to prevent the development of cracks.

These joints should be formed to provide a gap 1in. wide at the surface tapering to $\frac{3}{4}$ in. at the base, and before being primed and sealed with a bituminous sealing compound a suitable poison in liquid form such as pentachlorophenol should be poured into the joint to soak into the formation.

A membrane of poison should also be laid over the whole area of the formation supporting the slab. A typical lay-out of joints of this type is shown in Fig. 16. 6.[70]

CORROSION

This aspect must be given careful consideration in reinforced concrete structures because in addition to the greater likelihood of shrinkage cracks due to rapid drying, corrosion will develop rapidly when atmospheric temperatures are high.

Severe corrosion may be anticipated in coastal areas and reinforced concrete bridges and similar structures should be constructed in such a way as to avoid the development of cracks which may penetrate to the reinforcement. In order to achieve this it may be necessary to install more control joints than would normally be provided elsewhere and a greater depth of cover should be given to the reinforcement.

Particular care should be taken to ensure that all construction joints are properly prepared and if possible a groove should be formed at the surface so that a sealing compound may be applied to these joints if any cracks develop. Small triangular grooves at the surface effectively conceal construction joints and thus, apart from making provision for a sealing compound, they improve the general appearance of the structure.

OTHER CONSIDERATIONS

There are a number of other aspects which will vary in importance according to the particular conditions experienced on the site or in the area concerned. For example, it is quite common that due to

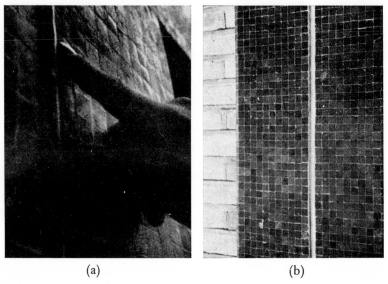

(a) (b)

FIG. 16. 7. Expansion and contraction effects at an expansion joint in the wall of a building. *Left :* Expansion, *right :* contraction.

shortages of supervisory staff, the remoteness of sites and lack of funds, the inspection and maintenance of structures such as bridges is carried out very infrequently.

In new construction also, problems may arise when inexperienced labour is introduced to new materials and methods of construction. Problems of this type, which include the effects of religious festivals such as Rammadan, may, and often do, affect nearly every aspect of the construction work.

The materials and methods chosen for jointing structures in tropical and sub-tropical climates and in under-developed areas therefore must not only be suitable from the point of view of performance, but they must be appropriate to the primitive methods of construction and storage which may exist on the site. Some of the more important safeguards which may be adopted in the choice of joint design and jointing materials are as follows :

DESIGN

The design and spacing of joints in structures must obviously be chosen to suit the conditions on the site. Careful planning of the structure in the design stage may be essential to ensure that the performance requirements of the jointing materials are, in fact, attainable. Under any circumstances the use of appropriate design methods to mollify the conditions to be accommodated by jointing materials will usually be found fully justified by subsequent economies and performance results.

In climates having extremes of temperature large volume changes of the concrete will occur. Typical variations incurring at joints in the walls of a building are illustrated in Fig. 16. 7. (a) and (b). The width of joints to be sealed with a sealing compound should, however, be kept to a minimum in order to reduce the tendency for the compound to flow or slump when exposed to high atmospheric temperatures.

It is recommended, therefore, that the width of joints be restricted to a maximum value of $\frac{1}{2}$in. and that joint spacings be chosen so that the movement at each joint is not likely to exceed $\frac{1}{10}$in. This could, in many climates, call for shrinkage or contraction joints at a spacing of 15ft. and expansion joints at a maximum spacing of 60ft.

It is suggested that these might be considered reasonable spacings for all structures in tropical or sub-tropical climates, particularly in view of the other conditions or circumstances which must be accommodated.

INSULATION

In climates having a large temperature range, methods of insulating concrete structures should be investigated in order to reduce the temperature variations in the concrete. Back-filling against walls with earth and the provision of an insulating screed to roofs are typical measures which may be adopted.

When a screed is laid on a roof, expansion joints should be formed immediately above those in the concrete, and it is desirable that a waterproof membrane be laid between the concrete and the screed, particularly when heavy rainfall may be expected during rainy seasons.

PROPERTIES OF JOINT FILLERS

There is no joint filler available at the present time which possesses the properties necessary to satisfy all the conditions to be anticipated in tropical and the sub-tropical climates.

Compressibility. The provision of expansion joints at frequent intervals is not always convenient or economical and even when such a measure is adopted it is often desirable for the joint filler to have very good compressibility and recovery after compression. The filler must, at the same time, be sufficiently rigid to permit the use of simple methods of installation and to withstand rough handling both on the site and in transit.

Flammability. In order to minimize fire risks during storage, which may be in the open (and due to misappropriation !), non-inflammable materials have advantages. Some of these materials, for example, certain foamed thermoplastics, tend to buckle when exposed to varying atmospheric conditions, thus complicating installation.

Durability. Mould growth and attack by insects are probably the most important destructive agencies affecting the life of joint fillers. A few joint fillers, such as synthetic rubber and thermoplastics, are virtually immune to attack, but other materials must be suitably impregnated. When an impregnant is used it should not be readily displaced by water because curing of the concrete is often effected by continual hosing down. Displacement of the impregnant due to this cause may lead to serious discoloration of the concrete adjacent to the joint in addition to reducing the durability of the joint filler.

SEALING COMPOUNDS

The most important property required in sealing compounds is usually low temperature susceptibility. This enables the compound to provide an effective seal without tending to flow in hot weather, or to lose adhesion or fracture when the concrete contracts rapidly. Rubber/bitumen compounds are among the most satisfactory materials available at present, but it is likely that the cold applied synthetic rubber compositions now being developed in the Northern Hemisphere may be even more suitable from the point of view of temperature accommodation.

These materials are not, however, easy to apply, and foolproof appliances should be employed to ensure correct apportioning of the components, mixing and application.

Storage. Compounds which are supplied in strip form, in particular, must be formulated and packed in such a way that they do not deteriorate when stored at high ambient temperatures for long periods or become deformed unduly during storage and removal from the pack. Clear instructions should be given by the manufacturers and site supervisors regarding the storage of these materials.

Quality. In general, it would appear that high quality sealing compounds are justified in order to cope with the difficult and often conflicting requirements encountered in many countries overseas. The opportunities for maintenance are often all too scarce because, in addition to remoteness, many structures must be in continuous use as soon as they are put into service. Moreover, the application of materials is usually more closely controlled during the construction period than conditions will allow subsequently.

In covered water-retaining structures, in particular, it is highly desirable that the joints in the walls and floors are sealed before the roof is constructed in order to avoid condensation difficulties.

Ease of application. It has already been indicated that ease of application is an important factor influencing the choice of a sealing compound. The performance of a sealing compound, and in particular the bond obtained with the concrete, is governed almost completely by the cleanliness of the joint and the care taken to ensure that the compound is heated and applied correctly.

When hot-applied sealing compounds are to be applied by inexperienced labout without close supervision, safe heating and application temperatures should not be critical ; a range of 20-30 deg. F. is desirable. The use of a primer, whilst being desirable if applied correctly, may on occasions prove to be an additional hazard because if it is given enough time to dry it may become coated with a layer of sand. If, on the other hand, it is not allowed to dry it will certainly impair adhesion.

The best solution to this problem is probably to ensure that the joints are clean and dry and omit the primer. Provided that the compound and the joints are properly prepared, satisfactory adhesion should be easily obtainable at high atmospheric temperatures.

WATERBARS

No special properties will, in general, be required in waterbars as far as performance in service is concerned. They should, however, be sufficiently robust to withstand rough handling on the site and in transit. They should not tend to deteriorate when stored at high atmospheric temperatures.

Rubber waterbars should be protected from direct sunlight as far as is practicable, and the jointing materials for the waterbar should have a good temperature accommodation.

On sites where the concrete is compacted by simple hand methods, waterbars having a simple profile are desirable. Consideration should also be given in the case of valve type waterbars to the depth of concrete cover provided to ensure the production of dense waterproof concrete around the valves.

A large number of difficult and often unpredictable conditions are frequently met in the construction and service life of concrete structures overseas. The development of cracks and the excessive movements occurring at joints in many structures which are already in existence point to the need for extreme caution in the design of structures and the selection and use of jointing materials.

The most desirable feature of all projected work is simplicity. The success of designs or construction methods, whether simple or complicated, must in practice rely on the intelligence, the experience and the conscientiousness of the labourers, skilled and semi-skilled men on the site.

BIBLIOGRAPHY

CHAPTER TWO

1. " Modern developments in reinforced concrete. Volume changes of concrete ", Constructional Review, April 1947, p. 29.
2. " Concrete engineering ", by J. Singleton-Green, Vol. II, 1935.
3. " The thermal expansion of concrete ", Bonnell and Harper, Journal of the Institution of Civil Engineers, 1950, Vol. 33, No. 4, pp. 320-330.
4. " Concrete roads, design and construction ", H.M.S.O., 1955, p. 166.
5. " Building materials, their elasticity and inelasticity ", Marcus Reiner, North Holland Publishing Company.
6. " Expansion joint practice in highway construction ", A. A. Anderson, American Society of Civil Engineers, Transactions, Vol. 114, 1949, paper No. 2384.
7. " The Structural Design of Concrete pavements ", L. W. Teller and E. C. Sutherland; Public Roads, Vol. 17, No. 9, p. 109.

CHAPTER THREE

8. " A Method of Studying the Formation of Cracks in a Material under Stress ", R. Jones, British Journal of Applied Physics, Vol. 3, No. 7, pp. 229-232.
9. " Strength and Plasticity of Concrete ", Ya. O. Berg, Doklady Akademii Nauk, U.S.S.R., Vol. 70, 1950, pp. 617-620.
10. " Tensile Strains in Concrete ", Parts I & II, F. A. Blakey and F. D. Beresford, C.S.I.R.O., Division of Building Research, Australia, 1953, Reports C2. 2-1, C2. 2-2.
11. " Symposium of the International Association of Testing and Research Laboratories for Materials and Structures ", A. M. Haas, D. Watstein and R. G. Mathey, Stockholm, June 1957.
12. " Die neue Theorie des Stahlbetons ", R. Saliger, 1950.
13. " Concrete in Motion ", F. A. Blakey, C.S.I.R.O. Division of Building Research, Australia, July 1962, Report L.12.
14. " Repairing Cracked Concrete with Plastics ", F. A. Blakey and E. N. Mattison, Building and Decorating Materials, Sydney, Vol. II, No. 1, pp. 36, 37.
15. " Use of Epoxy Resin for Bonding New Concrete to Old ", F. L. Burns, Research Investigation 60/15 State Rivers and Water Supply Commission, Victoria, Australia.
16,17." Failure and Repair of Concrete Structures ", S. Champion.
18. " Use of Epoxy Resin for Bonding New Concrete to Old ", F. L. Burns, Research Circular No. 6 State Rivers and Water Supply Commission, Victoria, Australia.

CHAPTER FIVE

19. " No-joint concrete pavement laid in Pennsylvania ". Ed. Mannix, Contractors and Engineers, July 1957, pp. 54-56.
20. " Concrete runways ", F. R. Martin. Pavings Development Group paper, October 1956.
21. " The Oxton by-pass extension ", R. A. Kidd. Institution of Civil Engineers, January 1956.

CHAPTER SIX

22. " Filling and sealing expansion joints ", J. T. Crandell, American Highways, January 1941, Vol. XX, No. 1, p. 21.
23. " Design and construction of concrete roads ", R. A. B. Smith and T. R. Grigson. Concrete Publications Limited, London 1946, p. 50.

Chapter Seven

24. " The refacing of four concrete dams ", W. G. M. Ternan, Civil Engineering and Public Works Review, February 1956, p. 197.
25. British Patent 640,055, " Improvements in or relating to the sealing of joints in concrete roads and the like ", R. S. Millard and P. L. Critchell.
26. " Some experiences with expansion joints in concrete pavements ", A. P. Anderson, Public Roads, Vol. 21, No. 3, pp. 57-61.
27. " Temporary sealing of joints in concrete roads on housing estates ", P. L. Critchell and R. S. Millard, Roads and Road Construction, August 1950.

Chapter Eight

28. " Expansion joints for concrete hydraulic structures ", H. T. Swanton, State Rivers and Water Supply Commission, Victoria, Australia; Constructional Review, December 1948.
29. " Rubber lining of a reservoir ", The Surveyor, 27th June, 1953.
30. " Rubber gas-main rings ", Bulletin of the Rubber Growers Association, February 1933.
31. " Development of articulation for large concrete canal structures ", H. G. Curtis, U.S. Bureau of Reclamation, Denver, Colorado; Western Construction News, May 1940.
32. " Rubber waterstops ", P. L. Critchell, Contractors Record and Municipal Engineering, 16th April, 1958.
33. " Bewegungsfugen ", 5th Edition, H. A. Kleinlogel, Wilhelm Ernst & John, Berlin, 1952.

Chapter Nine

34. " Repairs to a fractured expansion joint in a 10 M.G. reservoir ", R. G. Alcock, Journal of the Institution of Water Engineers, Vol. 8, No. 5, August 1954.
35. " Prestressed tanks at a sewage works ", Concrete and Constructional Engineering, Vol. LIII, No. 1, pp. 35-39.
36. " The design and construction of concrete roads overseas ", A. R. Collins and D. R. Sharp, Proceedings of the Institution of Civil Engineers, January 1958.
37. " Dowel-bar joints for airfield pavements ", J. A. Loe, Part II. Proceedings of the Institution of Civil Engineers, October 1952.
38. " The design of flat concrete roofs in relation to thermal effects ", Building Research Station Digest No. 12, H.M.S.O.
39. Provisional British Patent Specification No. 26291/57. " Improvements in floor strip ", T. Pooley.
40. " Sawn joints in concrete roads and airfield pavements ", L. S. Blake and L. O. Southey, The Surveyor, 14th April 1956.

Chapter Ten

41. " Concrete materials and practice ", L. J. Murdoch.

Chapter Eleven

42. " Concrete bridge design ", Portland Cement Association, Chicago, U.S.A.
42a " Expansion Joints in Bridges and Concrete Roads ", Waldemar Köster, trs. C. van Amerongen (M.Sc., A.M.I.C.E.) Maclaren and Sons.
43. " Control joints regulate effects of volume change in concrete masonry ", G. A. Mansfield, C. A. Sibrine and B. Wilk, Journal of the American Concrete Institute, Vol. 29, No. 1, July 1957.
44. " Weatherproof Joints Between Precast Concrete Panels ", D. Bishop. The Builder, 5th January 1962, 202 (6290), pp. 31-35.
45. " Shrinkage and temperature stresses in masonry ", R. E. Copeland, Journal of the American Concrete Institute, Volume 28, No. 8, February 1957.
46. " Long span prestressed concrete bridges constructed by the Freyssinet system ", Yves Guyon, Proc. Institution of Civil Engineers, Vol. 7, May 1957.
47. " Prestressed concrete roads ", J. P. Stott, Proc. Institution of Civil Engineers, Part II, October 1955.

BIBLIOGRAPHY

48. " Prestressed concrete cylindrical tanks ", L. R. Creasy, Proc. Institution of Civil Engineers, January 1958.

CHAPTER TWELVE

49. British Patent Application 30820/56, 10th October 1956. " Improvements relating to cutting tools ", S. Hill and J. F. Field-Smith.
50. " Resealing joints and cracks in concrete pavements (Minnesota) ", Bulletin No. 63, Highway Research Board.

CHAPTER THIRTEEN

51. " Concrete and cement mortar additives ", J. Singleton-Green and P. L. Critchell. Contractors Record and Municipal Engineering, 5th February 1958.
52. " Asphaltic bitumen roofing felts: their use in flat roof construction in tropical regions ", Colonial Building Note, No. 36, Building Research Station, D.S.I.R., Watford, August 1956.

CHAPTER FIFTEEN

53. Federal Specification HH-F-341A, 23rd November 1951. " Filler, expansion joint, preformed, non-extruding and resilient types (for concrete) ". Superintendent of Documents, United States Government Printing Office, Washington 25, D.C.
54. Federal Specification SS-R-406C, 18th July 1951: " Road and paving materials: methods of sampling and testing ".
55. Federal Specification SS-S-164, June 1956: " Sealer, hot-poured type for joints in concrete ".
56. " Standard methods for testing petroleum and its products ". Institute of Petroleum, 26, Portland Place, London, W.1.
57. Federal Specification SS-S-167: " Sealing compound, jet fuel resistant, hot applied, concrete paving ".
58. British Standard Specification 2499:1950. " Tests to assess the properties of concrete road joint sealing compounds ", H.M.S.O.
59. " The rheological properties of asphaltic bitumens ". J. P. Pfeiffer and P. M. van Doormaal. Inst. Petroleum Tech. Journal, Vol. 22, p. 152, 1936.
60. " Methodes d'Essai des produits pour joints de dalles en beton ", M. Duriez.
61. " Spécifications concernant les matières de scellement pour joints de revêtements en béton ". Centre de Recherches Routières, Bruxelles.
62. Deutscher Norm DIN1996: " Bitumen und Teer enthaltende Massen für Strassenbau und ähnliche Zwecke ". Beuth-Vertrieb GMBH, Berlin W15/Koln.
63. Federal Specification SS-S170: " Sealing compound, two-components, jet fuel resistant, cold-applied, concrete paving ".
64. Federal Specification TT-C-598: " Compound, caulking; plastic (for masonry and other structures) ".
65. " Specification for plastic caulking compound No. 1-GP-23-1946 " and " Schedule of methods of testing paints and pigments No. 1-GP-71 ". Canadian Government Purchasing Standards Committee, National Research Council, Ottawa, Canada.
66. British Standard Specification 903:1950: " Methods of testing vulcanized rubber ".

CHAPTER SIXTEEN

67. " Cracks in existing reinforced concrete buildings ", Kazuo-Ohno RILEM Symposium on bond and crack formation in reinforced concrete. Stockholm 1957.
68. " Practical considerations in the design and construction of concrete structures for waterworks ", by C. V. Green, Journal of the Institution of Water Engineers, Vol. 7, No. 6, October 1953.
69. " Foundations for building in the Orange Free State Goldfields ", J. E. Jennings, The Journal of the South African Institution of Civil Engineers, Vol. 49, Nos. 4 and 8, October 1950.
70. " Termite-proofing of buildings ", J. E. Jennings, Bulletin No. 4, National Building Research Institute, CSIR South Africa, May 1950.

CONVERSION TABLES

THE METRIC SYSTEM

Deci signifies 1/10	of the unit	
Centi ,, 1/100	,,	
Milli ,, 1/1,000	,,	
Deca ,,	10 times the unit	
Hekto ,, 100	,,	
Kilo ,, 1,000	,,	
Myria ,, 10,000	,,	

1 Inch	...	... =	2.54 Centimetres
1 Foot	...	... =	3.048 Decimetres
1 Yard	...	... =	9.144 Decimetres
1 Pole	...	... =	5.0292 Metres
1 Chain	...	... =	20.1168 Metres
1 Furlong	...	=	2.0116 Hektometres
1 Mile	...	... =	1.6092 Kilometres

Metric Conversion

Pounds per sq. in. × 0.0703 = Kilogrammes per sq. cent.   Kilos. per sq. cent. × 14.223 = Pounds per sq. in.

Conversion : Inches into Millimetres

Ins.	0	1	2	3	4	5	6	7	8	9	10	11
0	—	25.400	50.800	76.200	101.60	127.00	152.40	177.80	203.20	228.60	254.00	279.40
$\frac{1}{16}$	1.587	26.987	52.387	77.787	103.19	128.59	153.99	179.39	204.79	230.19	255.59	280.99
$\frac{1}{8}$	3.175	28.575	53.975	79.375	104.77	130.17	155.57	180.97	206.37	231.77	257.17	282.57
$\frac{3}{16}$	4.762	30.162	55.562	80.962	106.36	131.76	157.16	182.56	207.96	233.36	258.76	284.16
$\frac{1}{4}$	6.350	31.750	57.150	82.550	107.95	133.35	158.75	184.15	209.55	234.95	260.35	285.75
$\frac{5}{16}$	7.937	33.337	58.737	84.137	109.54	134.94	160.34	185.74	211.14	236.54	261.94	287.34
$\frac{3}{8}$	9.525	34.925	60.325	85.725	111.12	136.52	161.92	187.32	212.72	238.12	263.52	288.92
$\frac{7}{16}$	11.112	36.512	61.912	87.312	112.71	138.11	163.51	188.91	214.31	239.71	265.11	290.51
$\frac{1}{2}$	12.700	38.100	63.500	88.900	114.30	139.70	165.10	190.50	215.90	241.30	266.70	292.10
$\frac{9}{16}$	14.287	39.687	65.087	90.487	115.89	141.29	166.69	192.09	217.49	242.89	268.89	293.69
$\frac{5}{8}$	15.875	41.275	66.675	92.075	117.47	142.87	168.27	193.67	219.07	244.47	269.87	295.27
$\frac{11}{16}$	17.462	42.862	68.262	93.662	119.06	144.46	169.86	195.26	220.66	246.06	271.46	296.86
$\frac{3}{4}$	19.050	44.450	69.850	95.250	120.65	146.05	171.45	196.85	222.25	247.65	273.05	298.45
$\frac{13}{16}$	20.637	46.037	71.437	96.837	122.24	147.64	173.04	198.44	223.84	249.24	274.64	300.04
$\frac{7}{8}$	22.225	47.625	73.025	98.425	123.82	149.22	174.62	200.02	225.42	250.82	276.22	301.62
$\frac{15}{16}$	23.812	49.212	74.612	100.012	125.41	150.81	176.21	201.61	227.01	252.41	277.81	303.21

Conversion : Centimetres into Inches

Cms.	0	1	2	3	4	5	6	7	8	9
0	—	0.3937	0.7874	1.1811	1.5748	1.9685	2.3622	2.7559	3.1496	3.5433
10	3.9370	4.3307	4.7244	5.1181	5.5118	5.9055	6.2992	6.6929	7.0866	7.4803
20	7.8740	8.2677	8.6614	9.0551	9.4488	9.8425	10.236	10.630	11.025	11.417
30	11.811	12.205	12.598	12.992	13.386	13.780	14.173	14.567	14.961	15.354
40	15.748	16.142	16.535	16.929	17.323	17.717	18.110	18.504	18.898	19.291
50	19.685	20.079	20.473	20.866	21.260	21.654	22.047	22.441	22.835	23.228
60	23.622	24.016	24.410	24.803	25.197	25.591	25.984	26.378	26.772	27.165
70	27.559	27.953	28.347	28.740	29.134	29.528	29.921	30.315	30.709	31.102
80	31.496	31.890	32.284	32.677	33.071	33.464	33.858	34.252	34.646	35.039
90	35.433	35.827	36.221	36.614	37.008	37.402	37.795	38.189	38.583	38.976

Conversion : Feet into Metres

Feet	0	1	2	3	4	5	6	7	8	9
0	—	0.3048	0.6096	0.9144	1.2192	1.5240	1.8288	2.1336	2.4384	2.7432
10	3.0480	3.3528	3.6576	3.9624	4.2672	4.5720	4.8768	5.1816	5.4864	5.7912
20	6.0960	6.4008	6.7056	7.0104	7.3152	7.6200	7.9248	8.2296	8.5344	8.8392
30	9.1440	9.4488	9.7536	10.0584	10.3632	10.6680	10.9728	11.2776	11.5824	11.8872
40	12.1920	12.4968	12.8016	13.1064	13.4112	13.7160	14.0208	14.3256	14.6304	14.9352
50	15.2400	15.5448	15.8496	16.1544	16.4592	16.7640	17.0688	17.3736	17.6784	17.9832
60	18.2880	18.5928	18.8976	19.2024	19.5072	19.8120	20.1168	20.4216	20.7264	21.0312
70	21.3360	21.6408	21.9456	22.2504	22.5552	22.8600	23.1648	23.4696	23.7744	24.0792
80	24.3840	24.6888	24.9936	25.2984	25.6032	25.9080	26.2128	26.5176	26.8224	27.1272
90	27.4320	27.7368	28.0416	28.3464	28.6512	28.9560	29.2608	29.5656	29.8704	30.1752

Conversion : Metres into Feet

Metres	0	1	2	3	4	5	6	7	8	9
0	—	3.281	6.562	9.843	13.123	16.404	19.685	22.966	26.247	29.528
10	32.808	36.089	39.370	42.651	45.932	49.213	52.493	55.774	59.055	62.336
20	65.617	68.898	72.179	74.459	78.740	82.021	85.302	88.583	91.864	95.144
30	98.425	101.706	104.987	108.268	111.549	114.829	118.110	121.391	124.672	127.953
40	131.234	134.515	137.795	141.076	144.357	147.638	150.919	154.200	157.480	160.761
50	164.042	167.323	170.604	173.885	177.166	180.446	183.727	187.008	190.289	193.570
60	196.851	200.131	203.412	206.693	209.974	213.255	216.536	219.816	223.097	226.378
70	229.659	232.940	236.221	239.502	242.782	246.063	249.344	252.625	255.906	259.187
80	262.467	265.748	269.029	272.310	275.591	278.872	282.152	285.433	288.714	291.995
90	295.276	298.557	301.838	305.118	308.399	311.680	314.961	318.242	321.523	324.803

Square Measure

1 Square Inch	..	=	6.4516 Square Centimetres
1 Square Foot	..	=	9.2903 Square Decimetres
1 Square Yard	..	=	0.8361 Square Metres
1 Square Perch	..	=	25.292 Square Metres
1 Rood	..	=	10.1168 Square Decametres
1 Acre	..	=	40.4472 Square Decametres
1 Square Mile	..	=	2.5885 Square Kilometres

Cubic Measure

1 Cubic Inch	..	=	16.388 Cubic Centimetres
1 Cubic Foot	..	=	28.318 Cubic Decimetres
1 Cubic Yard	..	=	0.764 Cubic Metre

Liquid Measure

1 Gill	..	=	0.142 Litre
1 Pint	..	=	0.568 ,,
1 Quart	..	=	1.136 Litres
1 Gallon	..	=	4.546 ,,

Weight

1 Dram	..	=	1.772 Grammes
1 Ounce	..	=	28.352 Grammes
1 Pound	..	=	4.536 Hektogrammes
1 Stone	..	=	6.35 Kilogrammes
1 Quarter	..	=	12.7 Kilogrammes
1 Cwt.	..	=	50.8 Kilogrammes
1 Ton	..	=	1.016 Metric Tons

INDEX